Everyday Mathematics®

The University of Chicago School Mathematics Project

MY REFERENCE BOOK

Everyday Mathematics®

The University of Chicago School Mathematics Project

MY REFERENCE BOOK

McGraw Hill Education

Bothell, WA • Chicago, IL • Columbus, OH • New York, NY

The University of Chicago School Mathematics Project

Max Bell, Director, *Everyday Mathematics* First Edition
James McBride, Director, *Everyday Mathematics* Second Edition
Andy Isaacs, Director, *Everyday Mathematics* Third, CCSS, and Fourth Editions
Amy Dillard, Associate Director, *Everyday Mathematics* Third Edition
Rachel Malpass McCall, Associate Director, *Everyday Mathematics* CCSS and Fourth Editions
Mary Ellen Dairyko, Associate Director, *Everyday Mathematics* Fourth Edition

Authors
Mary Ellen Dairyko
James Flanders
Catherine Randall Kelso
Rachel Malpass McCall
Cheryl G. Moran

Writers
Sarah R. Burns
Judith Zawojewski

Digital Development Team
Carla Agard-Strickland, Leader
John Benson
Gregory Berns-Leone
Juan Camilo Acevedo
Scott Steketee

Technical Art
Diana Barrie, Senior Artist
Cherry Inthalangsy

UCSMP Editorial
Kristen Pasmore

Contributor
Jeanne Mills DiDomenico

www.everydaymath.com

Send all inquiries to:
McGraw-Hill Education
STEM Learning Solutions Center
8787 Orion Place
Columbus, OH 43240

ISBN: 978-0-02-138351-1
MHID: 0-02-138351-0

Printed in the United States of America.

2 3 4 5 6 7 8 9 QTN 19 18 17 16 15 14

Dear Children,

A reference book is a book that helps people find information. Some other reference books are dictionaries, encyclopedias, and cookbooks. *My Reference Book* can help you find out more about the mathematics you learn in class. You can read this reference book with your teacher, your family, and your classmates.

You will find lots of math in this book. Some of the things you will find are:

- counts
- number stories
- math tools
- fact strategies
- fact families
- addition strategies
- subtraction strategies
- measures
- clocks
- money
- shapes
- games

You will also find some big words. A grown-up can help you read these words.

We hope you enjoy this book.

Sincerely,
The Authors

Dear Family,

This book is a resource that your child can use to find out more about the mathematics learned in class. It is written for children to read with an adult.

You and your child can use this reference book to look up and review topics in mathematics to help with Home Links, to discuss the mathematics in the day's lesson, or to answer questions your child may have about mathematics in everyday life. You can explore these features with your child:

- A **Table of Contents** that lists the sections and shows how the book is organized
- **Essays** within each section, such as Number Stories, Ten Frames, Subtraction Fact Strategies, Adding Larger Numbers, and 2-Dimensional Shapes
- Opportunities for practice and discussion called **Try It Together**
- Two **photo essays** that show in words and pictures some ways that mathematics is used
- Directions for **games** that practice math skills
- An **index** to help you locate information

We hope you enjoy using this book with your child.

Sincerely,
The Authors

Standards for Mathematical Practice

Mazer Creative Services

What Are Mathematical Practices?

Read It Together

Mathematical practices are ways that mathematicians think and work as they do mathematics.

When you use mathematical practices, you are thinking and working like a mathematician.

These children are using mathematical practices.

In this section of *My Reference Book*, you will learn about eight mathematical practices.

Problem Solving: Make Sense and Keep Trying

Write a number sentence that has a sum between 21 and 25. Use two numbers from the box for the addends.

___ + ___ = A sum between 21 and 25

12	5	15	17	9	14

I need to find two numbers from the box that add up to a number between 21 and 25. My answer will be an addition number sentence.

Li

Make sense of your problem.
Li thinks about what she needs to do and what her answer will look like.

Hugo

Reflect on your thinking as you solve the problem.
Hugo starts by guessing. He learns from his guess that didn't work.

Keep trying when your problem is hard.
Hugo does not give up. He uses what he learns. He tries again when his first answer is wrong.

Check whether your answer makes sense.

On his second try, Hugo checks his sum to make sure it is between 21 and 25.

Grace finds the same answer another way. She draws a number line to show that $17 + 5 = 22$.

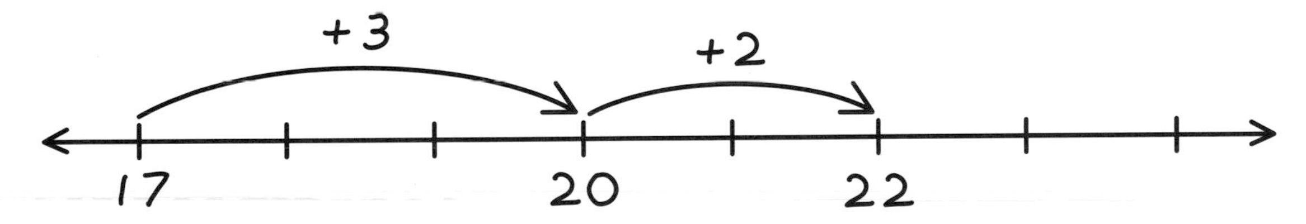

Solve problems in more than one way.

Hugo and Grace find the same answer in two different ways.

Compare strategies you and others use.

Hugo uses mental math. Grace uses a number line. Both strategies help them find answers.

Mathematical Practice 1: Make sense of problems and persevere in solving them.

Try It Together

Find another way to solve the problem. Tell how you know your answer makes sense.

Create and Make Sense of Representations

75 + 7 = ?

Dell represents represents 75 + 7 on a number line.

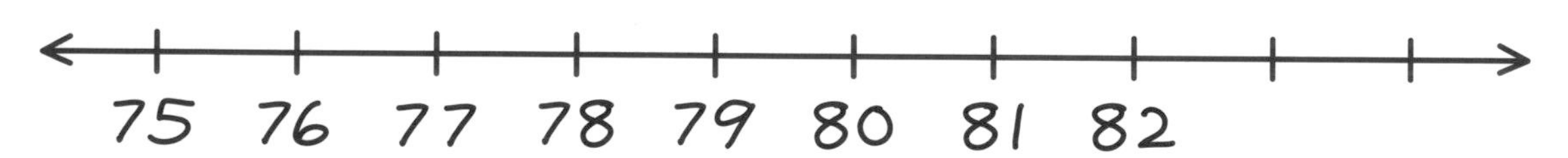

Dell

Create mathematical representations using numbers, words, pictures, symbols, gestures, tables, graphs, and concrete objects.

Dell draws a number line and uses it to add 75 + 7. He also explains what he did in words.

Grace represents 75 + 7 on a number grid.

71	72	73	74	(75)	76	77	78	79	80
81	(82)	83	84	85	86	87	88	89	90

Grace

Make connections between representations.

Grace notices that she and Dell both counted by 1s to solve the problem, but she used a number grid, and Dell used a number line.

Hugo represents 75 + 7 with base-10 blocks.

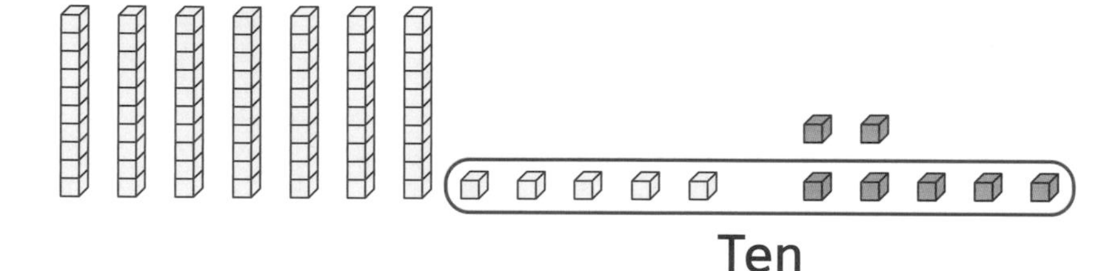

75 + 7 = 82

I show 7 tens and 5 ones and then another 7 ones. I make another ten by putting together 5 ones and 5 ones. So 7 tens and 1 ten are 8 tens. I have 2 ones left. That makes 8 tens and 2 ones, or 82. I used base-10 blocks and got the same answer as Grace and Dell.

Hugo

Make sense of the representations you and others use.

Hugo notices that the number grid, the number line, and the base-10 blocks all show that 75 + 7 = 82. All the children are using representations they understand.

Mathematical Practice 2: Reason abstractly and quantitatively.

Try It Together

How is adding on a number line like adding on a number grid?

How is adding with base-10 blocks different from adding on a number line or a number grid?

Make Sense of Others' Thinking

Use all 3 of these clues. Name a number that is:

- Greater than 5
- Less than 13
- A sum of doubles

Make mathematical conjectures and arguments.
Dell makes a ***conjecture*** *when he guesses 9. He uses mathematical thinking and what he knows from the problem. Dell's guess comes from two of the clues.*

Mason makes an argument that the answer cannot be 9 because it is not a sum of doubles.

Make sense of others' mathematical thinking.
Dell makes sense of Mason's argument and figures out a better answer. Dell's answer now fits all the clues.

Mathematical Practice 3: Construct viable arguments and critique the reasoning of others.

Try It Together

What is another number that fits all the clues? Use math to make a good guess, or conjecture. Find a partner and make an argument to show you are correct.

Find a partner who has a different answer. Make an argument that your partner's answer is or is not correct.

Make Models to Solve Problems

Rosa and Li want to know how many pets live in their neighborhood.

Rosa gathers data with a tally chart. She makes a tally mark for each pet in the neighborhood.

Pets	Tallies
Dogs	卌
Cats	卌 //
Rabbits	//

Rosa

5 dogs + 7 cats + 2 rabbits = ? pets

5 + 7 + 2 = 14 pets

Use mathematical models to solve problems and answer questions.

Rosa uses tallies to keep track of the different pets. She uses a number model to find how many pets there are in all.

Li gathers data with a graph. She draws an X for each pet in the neighborhood.

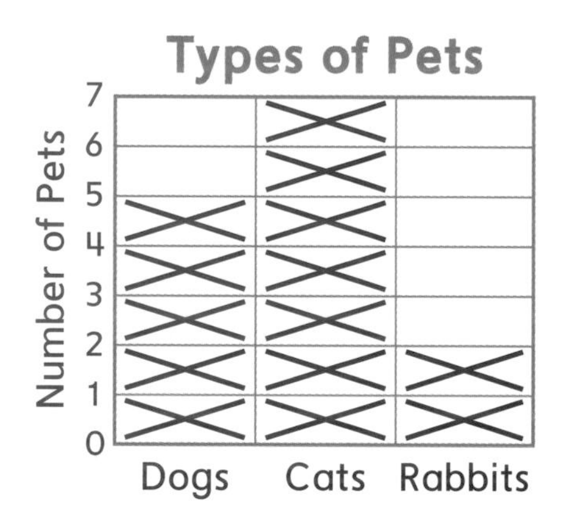

Model real-world situations using graphs, drawings, tables, symbols, numbers, diagrams, and other representations.

Rosa uses tallies and a number model to show the number of pets. Li uses a graph.

Pets	Tallies
Dogs	卌
Cats	卌 //
Rabbits	//

Types of Pets

Number of Pets

7
6
5
4
3
2
1
0

Dogs Cats Rabbits

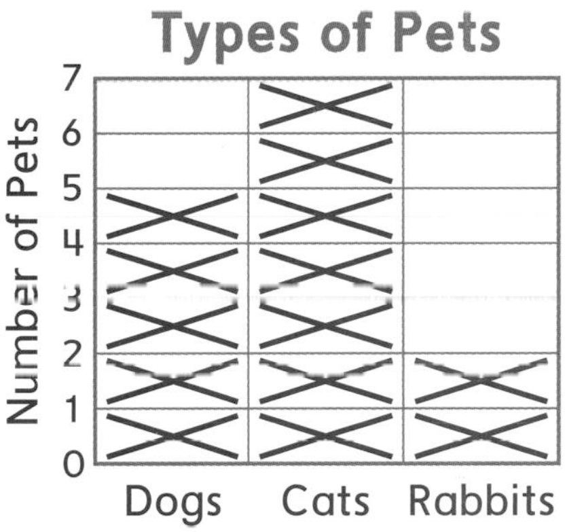

Mathematical Practice 4: Model with mathematics.

Try It Together

Use a drawing to model the number of pets in the neighborhood.

Choose Tools to Solve Problems

17 + 20 = ?

I started by putting out 17 counters. But using all those counters would take too long, and it's easy to make a mistake. So, I will try base-10 blocks instead.

Emma

Choose appropriate tools.

Emma decides that base-10 blocks are better than counters for this problem.

Emma shows 17 and 20 with base-10 blocks.

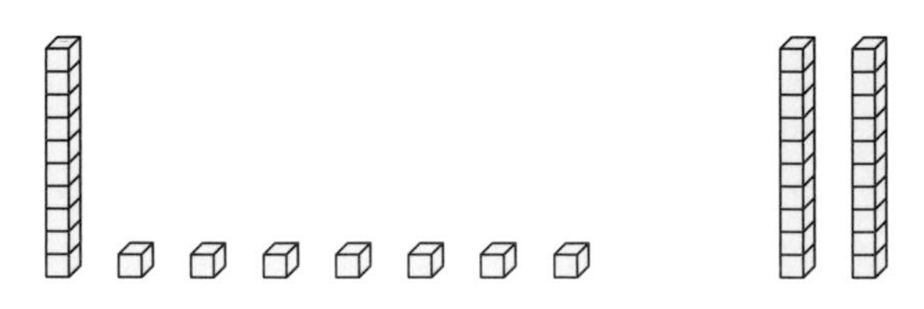

1 ten + 2 tens = 3 tens.

There are 3 tens and 7 ones in all. The sum is 37.

To check her answer, Emma uses a number grid to add 17 + 20.

11	12	13	14	15	16	17	18	19	20
21	22	23	24	25	26	27	28	29	30
31	32	33	34	35	36	37	38	39	40

Emma starts at 17 and moves down 2 rows to add 2 tens. She lands on 37, the same answer she got with base-10 blocks.

Use tools efficiently and make sense of your results.
Emma uses base-10 blocks to find the answer. She uses the number grid to check her answer.

Mathematical Practice 5: Use appropriate tools strategically.

Try It Together

What tool would you use to solve the problem? Explain how you would use it.

Be Careful and Accurate

14 stickers + 20 stickers = ? stickers

Dell told his teacher that he added 20 to 14 and got 34. Dell's teacher asks him to explain his thinking more clearly.

Explain your mathematical thinking clearly and precisely.
Dell uses details to tell how he added 14 and 20.

Use clear labels, units, and mathematical language.
Dell uses clear math language when he says, "20 is 2 tens." His answer uses the unit "stickers."

Is Ruler A or B better for measuring the length of a pencil? Explain.

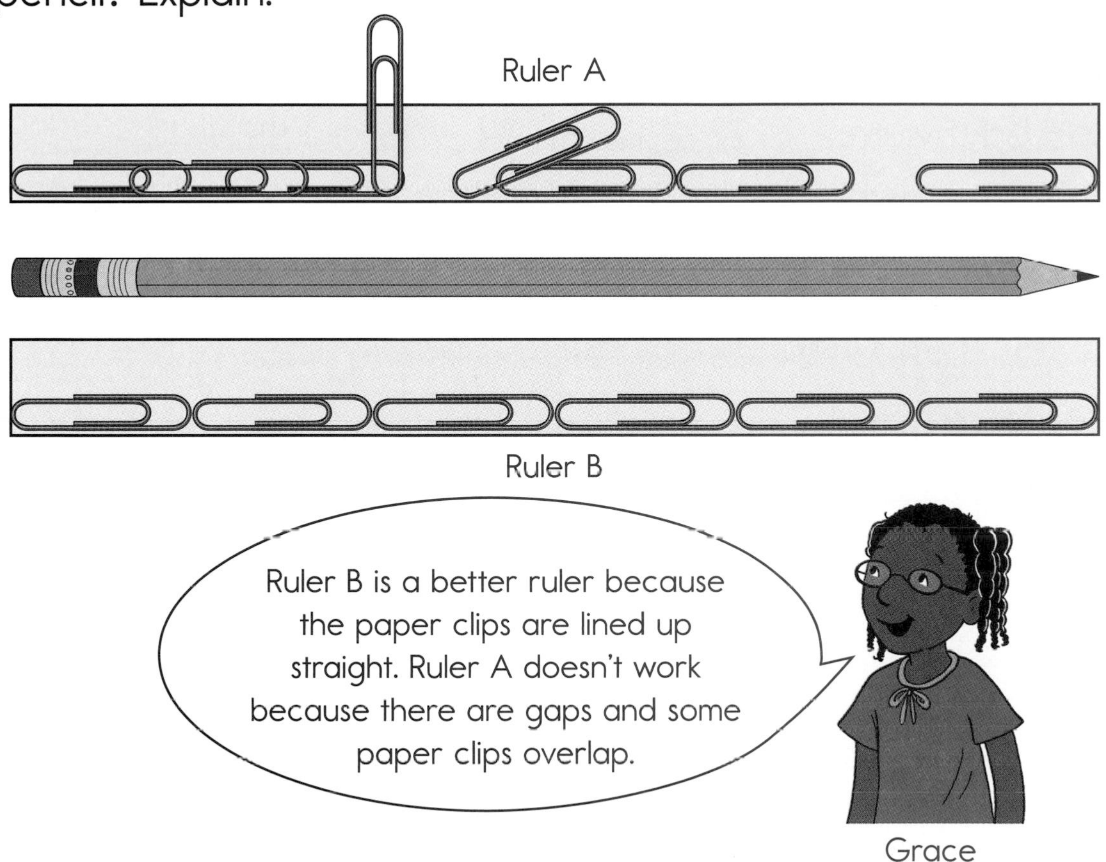

Think about accuracy and efficiency when you count, measure, and calculate.
Grace sees that Ruler B would give a more accurate, or correct, measurement than Ruler A.

Use an appropriate level of precision for your problem.
Paper clips are a good choice for measuring the length of a pencil because they are the right size for the job.

Mathematical Practice 6: Attend to precision.

Find and Use Patterns and Properties

Find each sum: 22 + 10 34 + 10 49 + 10

Mason counts by 1s on the number grid to solve these problems.

21	22	23	24	25	26	27	28	29	30
31	32	33	34	35	36	37	38	39	40
41	42	43	44	45	46	47	48	49	50
51	52	53	54	55	56	57	58	59	60

22 + 10 = 32

34 + 10 = 44

49 + 10 = 59

Mason

Look for mathematical structures such as categories, patterns, and properties.

Mason sees a pattern for adding 10 to numbers on the number grid.

Emma uses the turn-around rule for adding 3 + 8 on a number line.

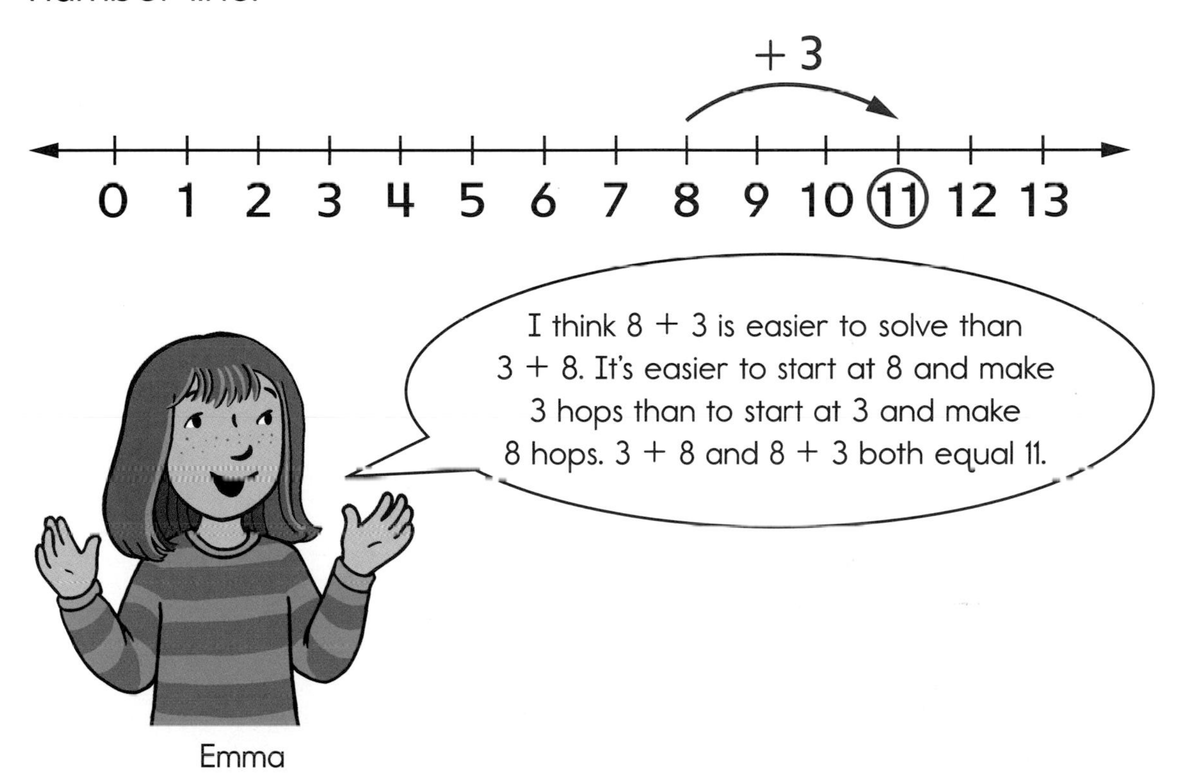

Emma

Use structures to solve problems and answer questions.
Emma uses the turn-around rule, or turn-around property, to find the answer in an easier way.

Mathematical Practice 7: Look for and make use of structure.

Try It Together

Which of these problems would be easier to solve using the turn-around rule? Explain.

2 + 9 = ?   7 + 4 = ?   3 + 6 = ?

Create Rules and Shortcuts

Rosa knows her doubles facts. She knows that

$7 + 7 = 14$ $8 + 8 = 16$ $9 + 9 = 18$

Rosa figures out how to use doubles to add near doubles:

$7 + 8 = ?$ $8 + 9 = ?$

If $7 + 7 = 14$, then $7 + 8$ is 1 more.
So, $7 + 8 = 15$. I know $8 + 8 = 16$,
so $8 + 9$ is 1 more. So, $8 + 9 = 17$.

Rosa

Create and justify rules, shortcuts, and generalizations. *Rosa justifies, or explains, how she uses doubles to help her figure out other facts.*

Mathematical Practice 8: Look for and express regularity in repeated reasoning.

Try It Together

If you know $10 + 10 = 20$, what is $10 + 11$?

If you know $25 + 25 = 50$, what is $25 + 26$?

Operations and Algebraic Thinking

Number Stories

Read It Together

Number stories are stories that use numbers. You can solve number stories using diagrams, drawings, number lines, and other tools and strategies.

Some number stories are about a total and its parts. You can use a **parts-and-total diagram** to help you solve them.

There are 8 yellow crayons and 6 blue crayons. How many crayons are there in all?

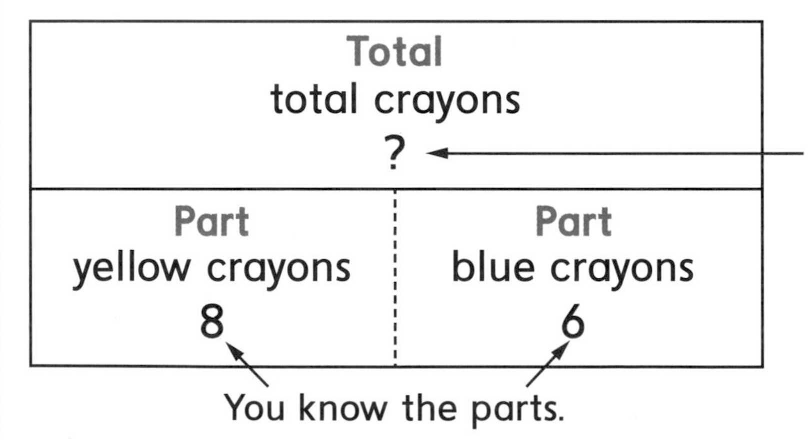

Number model: $8 + 6 = ?$
Add the parts to find the total.
Number model: $8 + 6 = 14$
There are **14** crayons in all.

Sometimes you need to find one of the parts.

There are 20 children on a bus.
9 children are girls. How many children are boys?

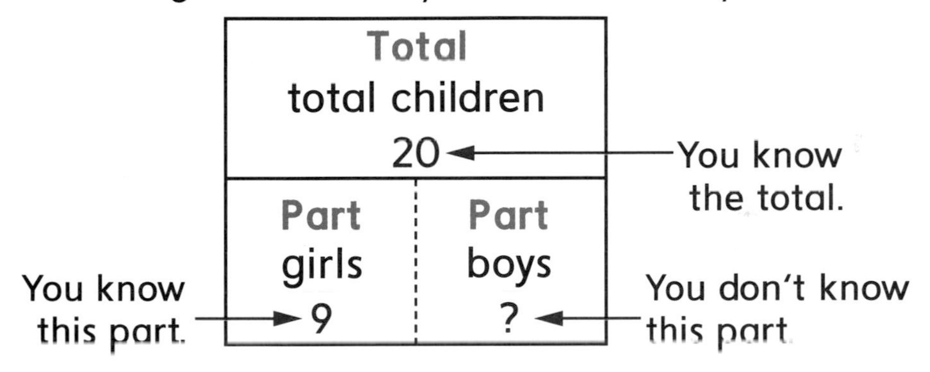

Number model: 20 − 9 = **?**

You can subtract to solve the problem.
Subtract the part you know from the total.
The answer is the other part.

Number model: 20 − 9 = **11** **11** boys on the bus

You can also count up to solve the problem.
Start with the part you know. Count up to the total.
The amount you count up is the other part.

Number model: 9 + **?** = 20

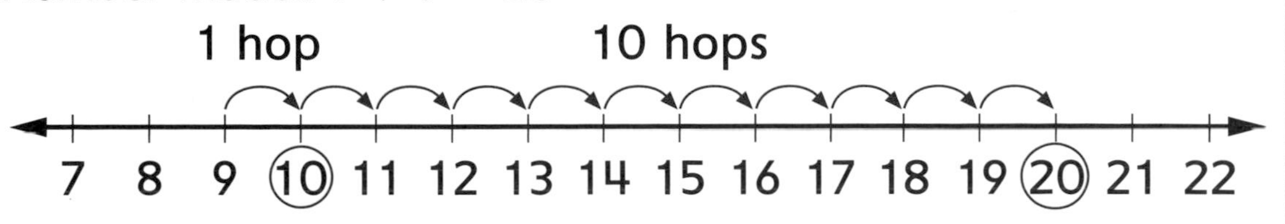

1 + 10 = **11**

Number model: 9 + **11** = 20 **11** boys on the bus

You can make a drawing to help you solve a number story about a total and its parts.

Together Ethan and his sister, Ava, have 14 fish. Ethan has 8 fish. How many fish does Ava have?

Number model: 8 + ☐ = 14

E E E E E E E
1 2 3 4 5 6 7

E A A A A A A
8 1 2 3 4 5 6

Name the parts and the total for this number story.

Number model: 8 + **6** = 14

Ava has **6** fish.

Try It Together

Write a number story for this number model:

5 + ☐ = 13

Some number stories are about a change. The number you start with changes to more or changes to less. You can use a **change diagram** to help you solve these stories.

Britney had 7 shells.
She found 9 more shells.
How many shells does Britney have in all?

This is a change-to-more story.

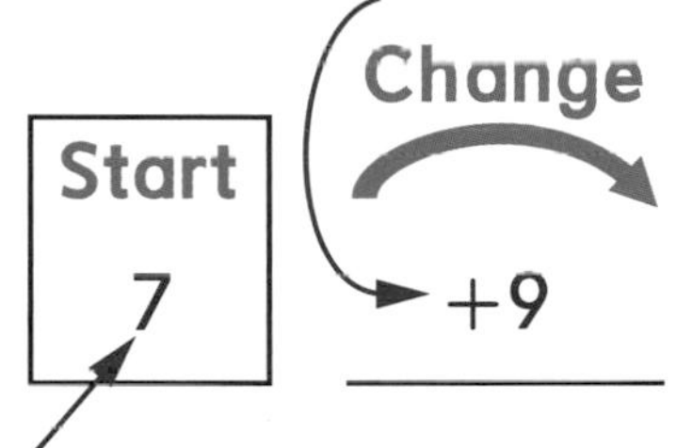

Number model: $7 + 9 = $ **?**

Number model: $7 + 9 = $ **16**

16 shells in all

You can use other tools and strategies to solve change stories.

There are 15 children on a bus.
6 children get off the bus.
How many children are left on the bus?

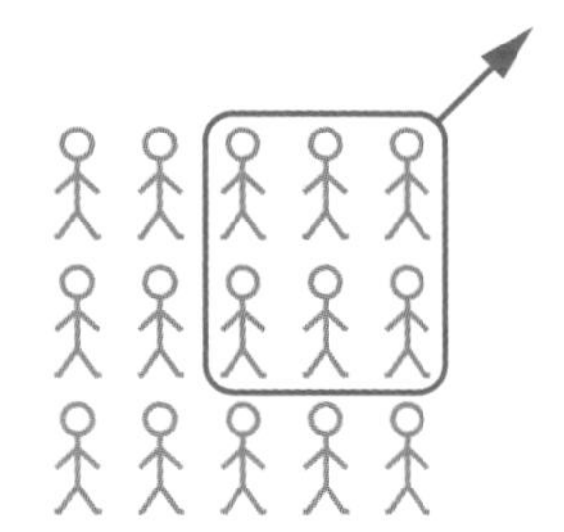

This is a change-to-less story.
You can use a drawing.

In this story, you don't know the ending number or how many children are still on the bus.

Number model: 15 – 6 = **?**
Number model: 15 – 6 = **9**
There are **9** children left on the bus.

You can use a number line to solve the problem.

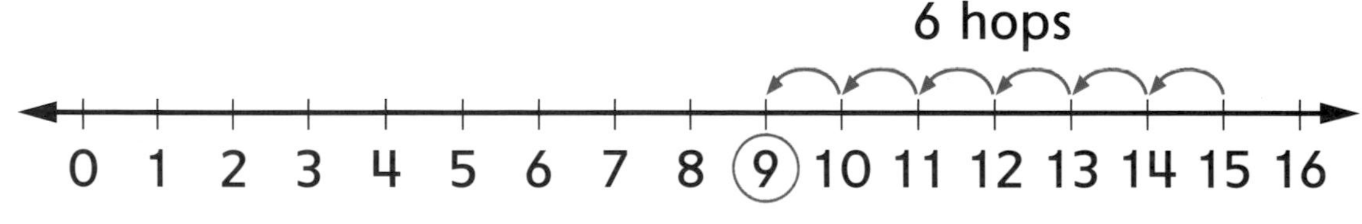

Number model: 15 – 6 = **9** children

Sometimes you need to find the change in a change story.

The morning temperature was 20°F.
The afternoon temperature was 32°F.
What was the temperature change?

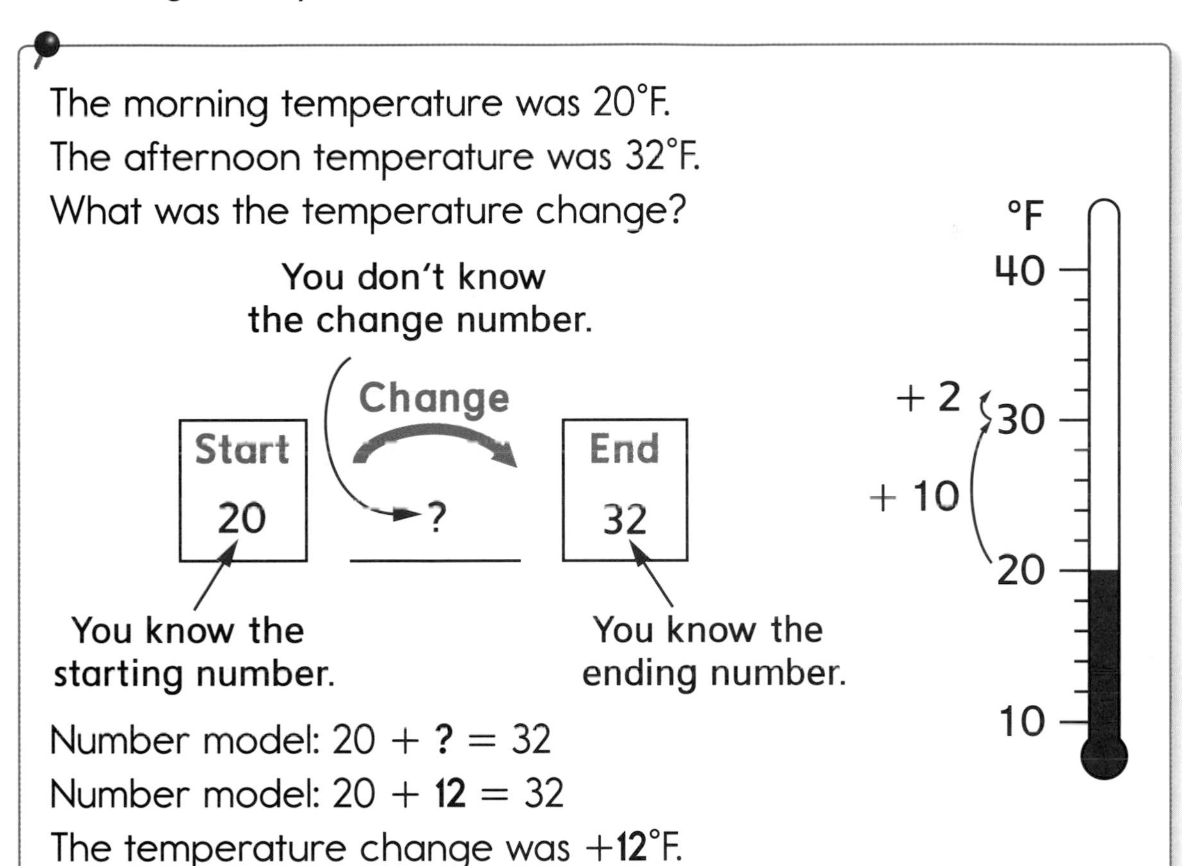

Number model: 20 + **?** = 32

Number model: 20 + **12** = 32

The temperature change was +**12**°F.

Try It Together

Take turns with a partner making up and solving number stories.

Some number stories are about comparisons. You can use a **comparison diagram** to help you solve them.

Jim is 13 years old. Ron is 9 years old.
How many years older is Jim than Ron?

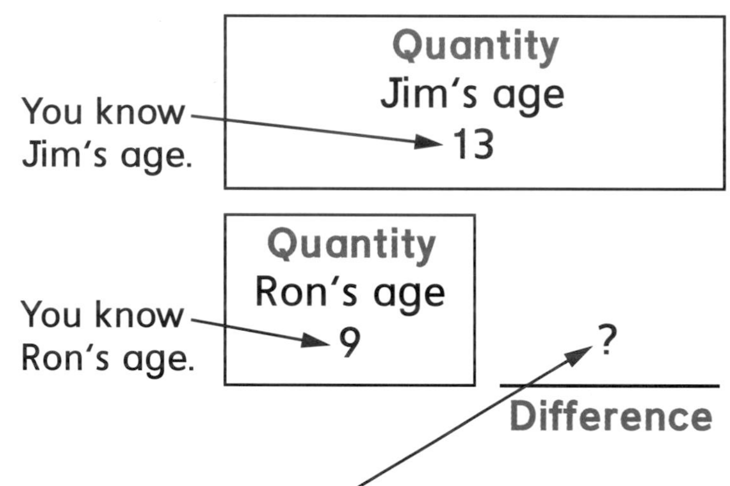

Number model: $13 - 9 =$ **?**

You can subtract to find the **difference.**

Start with the larger number.

Subtract the smaller number.

The answer is the difference.

Number model: $13 - 9 =$ **4**

Jim is **4** years older than Ron.

You can also count up to find the difference.
Start with the smaller number.
Count up to the larger number.
The amount you count up is the difference.

Number model: 9 + **?** = 13

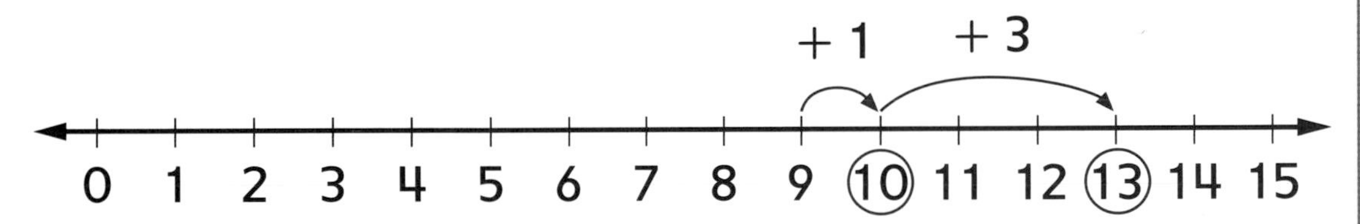

1 + 3 = **4**
Number model: 9 + **4** = 13
Jim is **4** years older than Ron.

Try It Together

Tell your partner a comparison number story.

You can use different strategies to help you solve a number story about groups with equal numbers of objects.

Mia has 4 packs of crayons.
There are 5 crayons in each pack.
How many crayons are there in all?

You can write different number models.

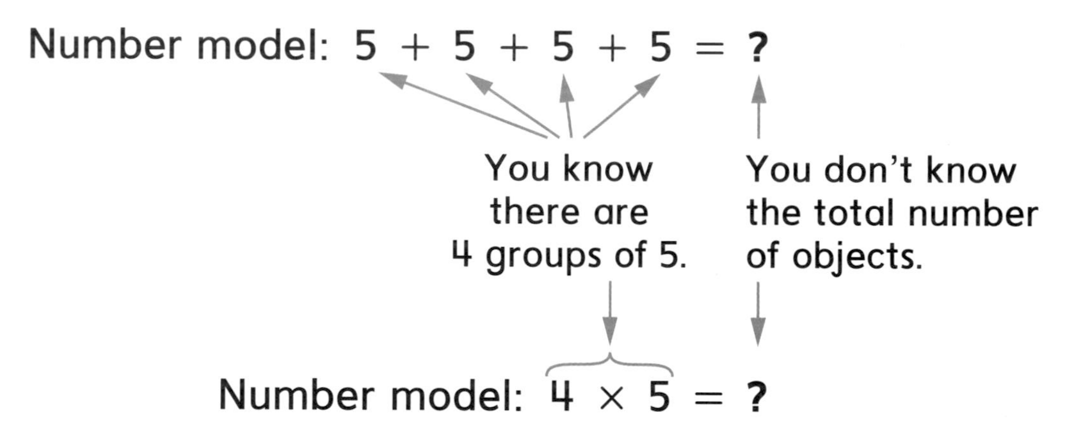

There are many ways to solve the problem.

You can draw a picture.

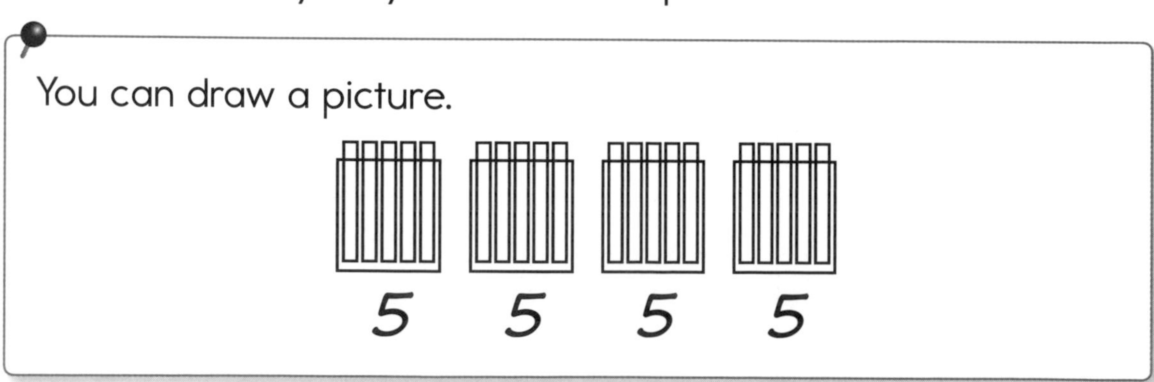

You can use an arrangement of dots in rows and columns called an **array.**

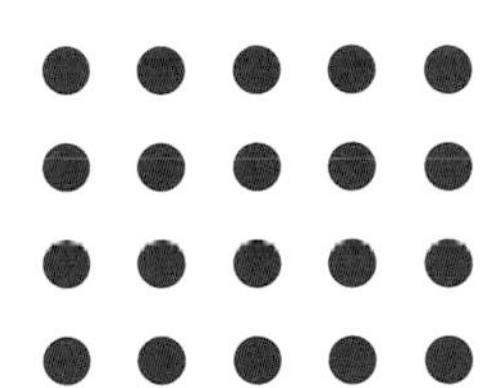

4 rows of 5 dots represent 4 packs of 5 crayons.

You can skip count by 5s.

5, 10, 15, **20**

You can add 5 four times.

Number model: $5 + 5 + 5 + 5 = \mathbf{20}$

4 groups of 5 is **20** altogether.

There are **20** crayons in all.

Another way to write this is $4 \times 5 = \mathbf{20}$.

Two-Step Number Stories

Sometimes you need to use more than 1 step to solve a number story. You can use different tools and strategies to help you.

On Monday, you had 12 papers in your desk at school. On Tuesday, you took 5 papers home.

On Wednesday, you put 9 more papers in your desk. How many papers are in your desk now?

You can make a drawing.

Monday, Tuesday   Wednesday

16 papers now

You can use a number line.

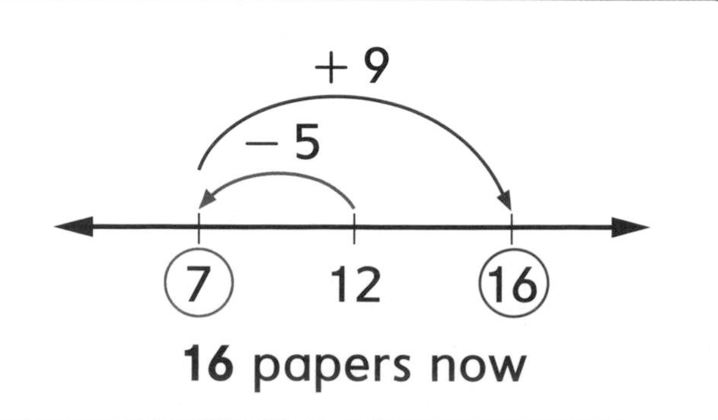

You can use diagrams.

Start	Change	End
12	− 5	?

Number model: 12 − 5 = 7.
7 papers at the end of Tuesday.

Start	**Change**	**End**
7	+ 9	?

Number model: 7 + 9 = 16
16 papers now.

You can use number models.

12 − 5 = **7**

7 + 9 = **16**

Now there are **16** papers.

Try It Together

Solve the number story below in two different ways.
How are the ways alike? How are they different?

Ellie has 15 yellow blocks and 25 blue blocks. She gives 12 blocks to Tommy. How many blocks does she have now?

Addition and Subtraction

Read It Together

We use words and symbols for addition and subtraction.

Use + to show addition.

$5 + 4 = \mathbf{9}$

$$\begin{array}{r} 5 \\ +\ 4 \\ \hline \mathbf{9} \end{array}$$

addend

sum

5 pennies and 4 more pennies is **9** pennies.

Use – to show subtraction.

$9 - 4 = \mathbf{5}$

$$\begin{array}{r} 9 \\ -\ 4 \\ \hline \mathbf{5} \end{array}$$

difference

9 pennies take away 4 pennies equals **5** pennies.

A **number line** can help you solve addition problems.

$5 + 4 = ?$

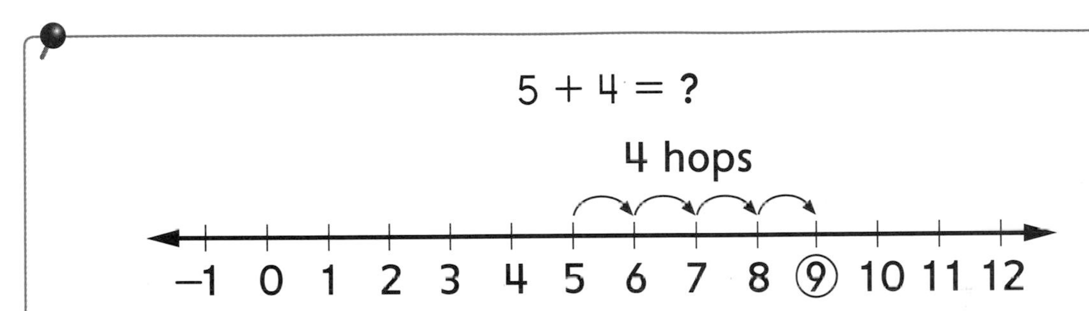

1. Start at 5.
2. **Count up** 4 hops.
   Land on **9.**

$5 + 4 = 9$

$6 + \square = 15$

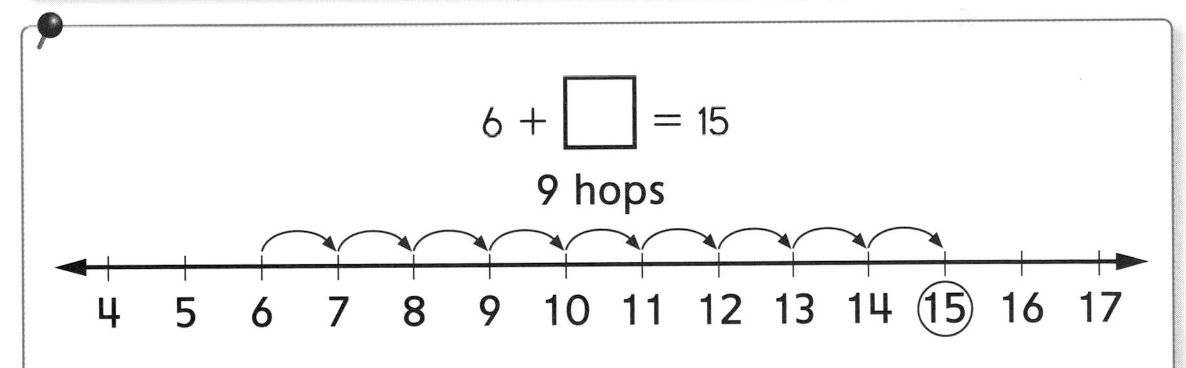

1. Start at 6.
   Think: How many hops from 6 to 15?
2. **Count up** to 15.
   There are 9 hops.

$6 + 9 = 15$

A number line can help you solve subtraction problems.

$$10 - 7 = ?$$

One way:

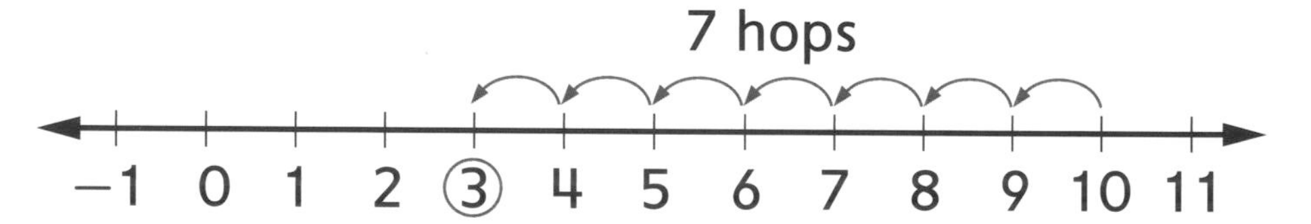

1. Start at 10.
2. **Count back** 7 hops. Land on 3.

$$10 - 7 = \mathbf{3}$$

Another way:

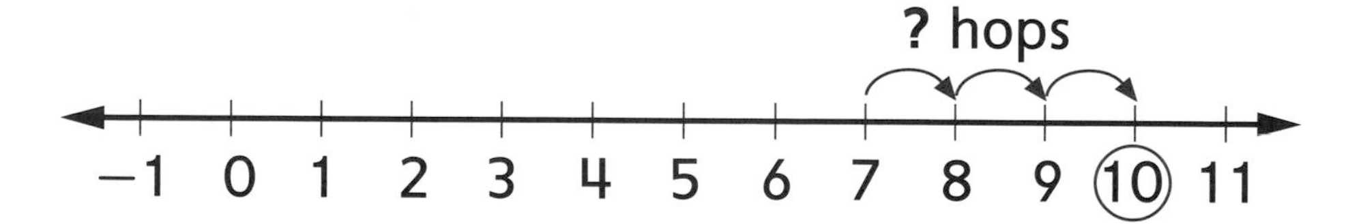

1. Start at 7.
   Think: How many hops from 7 to 10?
2. **Count up** to 10. There are 3 hops.

$$10 - 7 = \mathbf{3}$$

Ten Frames

A **ten frame** is a tool to help you think about numbers. It has 2 columns with 5 squares in each column. It has a total of 10 squares.

This ten frame shows 8, but you can see the 8 in different ways:

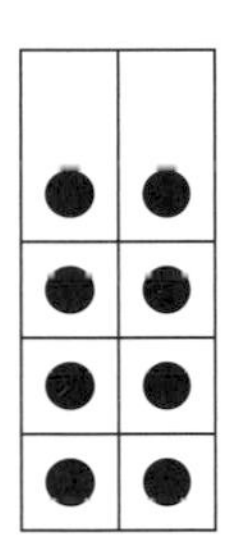

- You can see 2 rows of 4.
  $4 + 4 = 8$
- You can count by 2s.
  2, 4, 6, 8
- You can see 10 with 2 missing.
  $10 - 2 = 8$

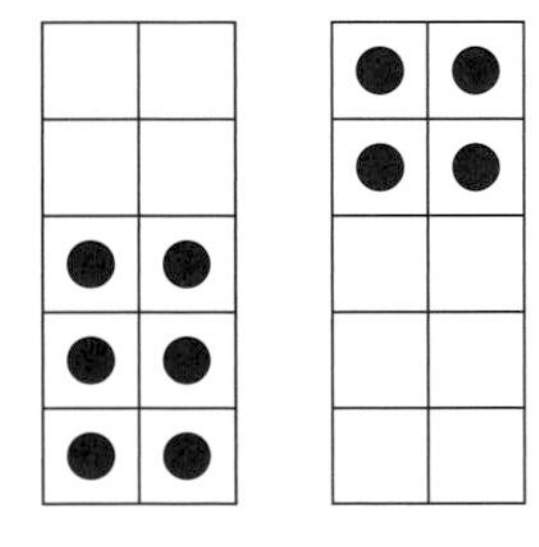

A **double ten frame** is made of two ten frames. It can help you think about larger numbers and adding numbers together.

Addition Helper Facts

Helper facts are facts you know well. They can help you figure out facts that you do not know.

Doubles are addition facts that have the same number for both addends, such as $6 + 6 = 12$ and $5 + 5 = 10$. You can think of doubles in different ways. We see many doubles in everyday life.

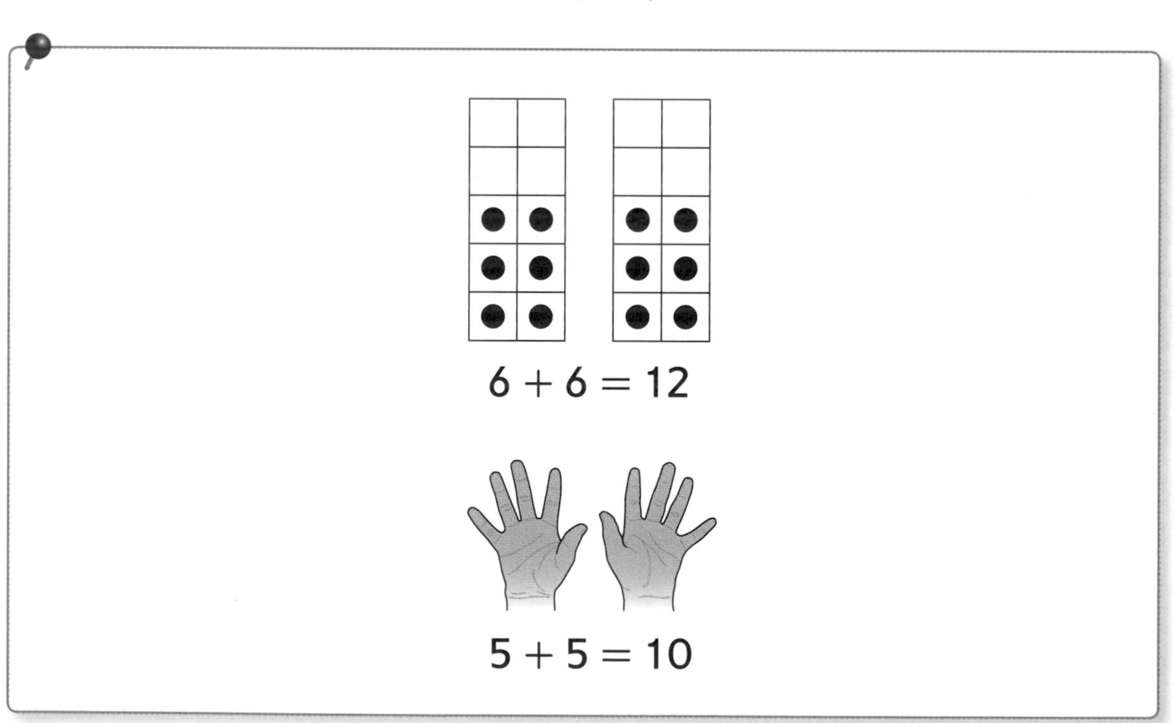

Combinations of 10 are addition facts with numbers that add to 10, such as $6 + 4 = 10$ and $2 + 8 = 10$. Combinations of 10 can be helper facts, too.

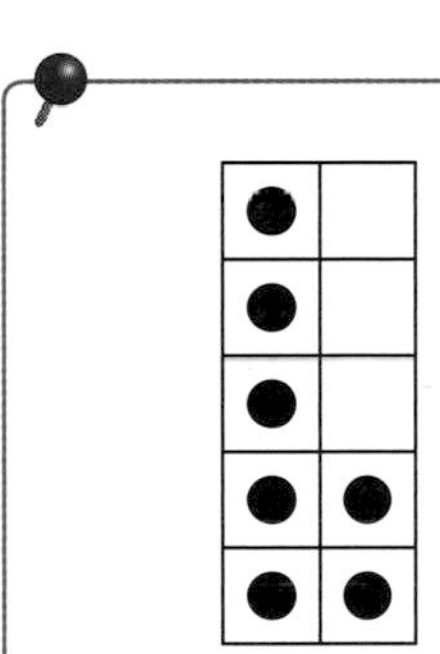

I see 7 dots and 3 empty spaces.
I see $7 + 3 = 10$.
I have a combination of 10.

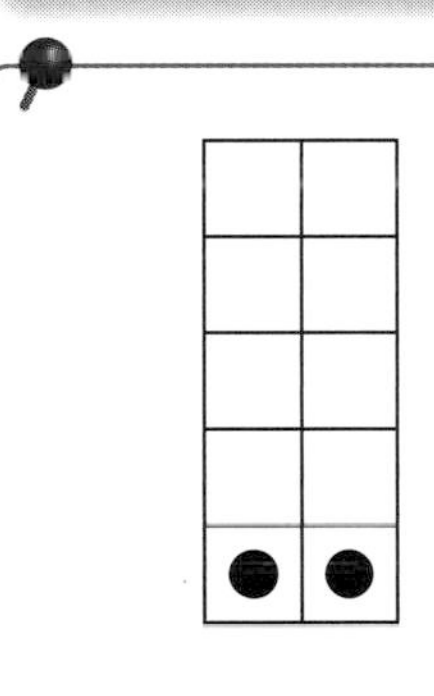

I see 2 dots in the bottom row and 4 rows of 2 empty spaces. That's 8 empty spaces. I see $2 + 8 = 10$.
I have another combination of 10.

Try It Together

Make lists of all the doubles facts and combinations of 10.

Strategies to Make Addition Easier

You can use **near doubles** to solve facts that are close to a double.

$4 + 5 = ?$

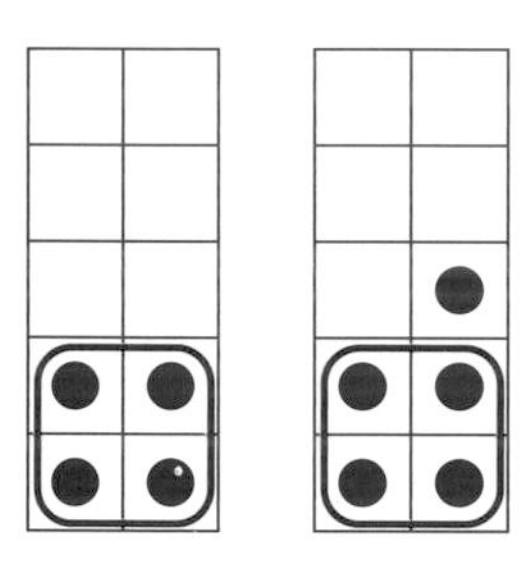

You can see the helper doubles fact $4 + 4 = 8$. Since there is 1 extra dot, you add 1 more and get 9.

$4 + 4 + 1 = 9$

$4 + 5 = \mathbf{9}$

Using near doubles works well when the two addends are close together.

$5 + 7 = ?$

Think: $5 + 5 = 10$, so $5 + 5 + 2 = 12$.

$5 + 7 = \mathbf{12}$

Try It Together

Sort through your fact triangles and find facts you could solve using near doubles.

Sometimes it is easier to break apart one of the addends to make a combination of 10 with the other addend.

This is called **making 10.**

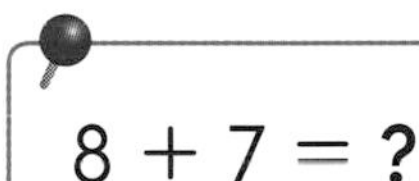

$8 + 7 = ?$

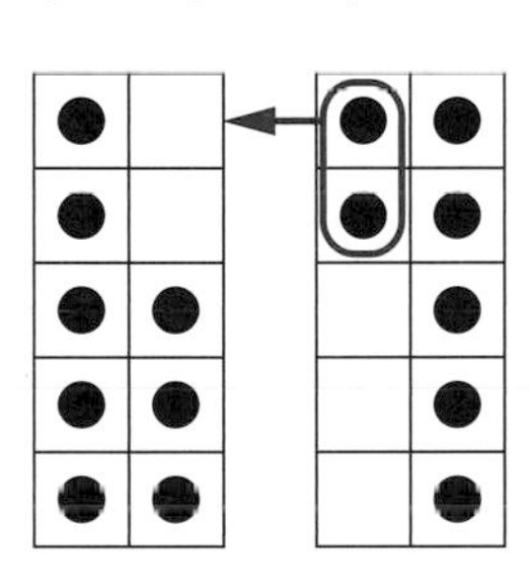

You can move 2 dots over to make 10 because $8 + 2 = 10$.
There are 5 more dots.
$10 + 5 = 15$.
So $8 + 7 =$ **15.**

Making 10 works well with larger addition facts and when one of the addends is close to 10.

$9 + 6 = ?$
Think: $9 + 1 = 10$, so $9 + 1 + 5 = 15$.
$9 + 6 =$ **15**

Try It Together

Sort through your fact triangles and find facts you could solve by making 10.

The **turn-around rule** says you can add two numbers in either order. Sometimes changing the order makes it easier to solve problems.

If you don't know what $3 + 8$ is, you can use the turn-around rule to help you, and solve $8 + 3$ instead. $8 + 3$ is easy to solve by counting on.

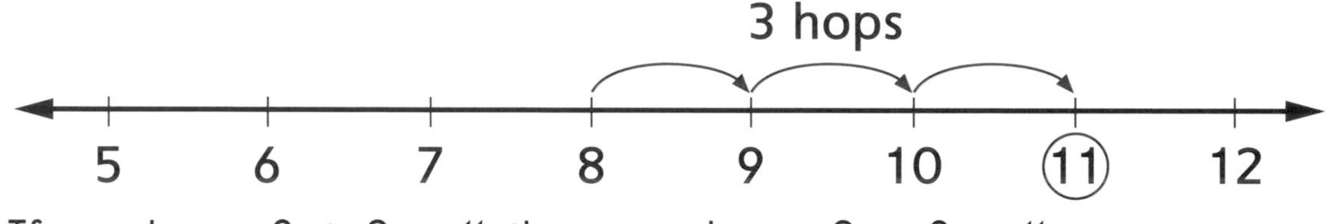

If you know $8 + 3 = 11$, then you know $3 + 8 = 11$.

Grouping addends together in different ways can make adding easier.

$4 + 9 + 1 = ?$

It can be easier to start with $9 + 1 = 10$.

$4 + 9 + 1 = ?$

$4 + 10 = \mathbf{14}$

$3 + 6 + 6 = ?$

It can be easier to start with the double $6 + 6 = 12$ then add 3 more to get 15.

$3 + 6 + 6 = ?$

$3 + 12 = \mathbf{15}$

When you add 0 to a number, you are not adding anything more, so the number stays the same.

$2 + 0 = 2$

$5 + 0 = 5$

When you add 1 to a number, the sum is the next number you say when you count by 1s.

$2 + 1 = 3$

$5 + 1 - 6$

Try It Together

Solve.

$6 + 7 = ?$ $\quad$ $25 + 0 = ?$ $\quad$ $4 + 8 = ?$

Choose one problem. Explain to a partner how to use a strategy to solve it.

Fact Families

Read It Together

A **fact family** is a group of related facts that uses the same numbers.

Dominoes can help you find fact families.

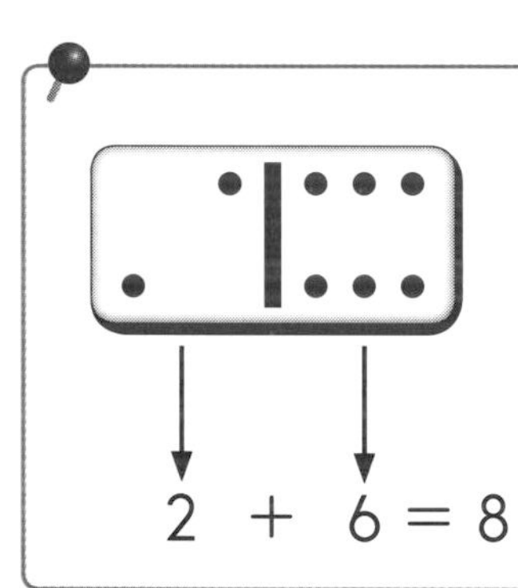

$2 + 6 = 8 \quad 8 - 6 = 2$

$6 + 2 = 8 \quad 8 - 2 = 6$

This is the fact family for 2, 6, and 8.

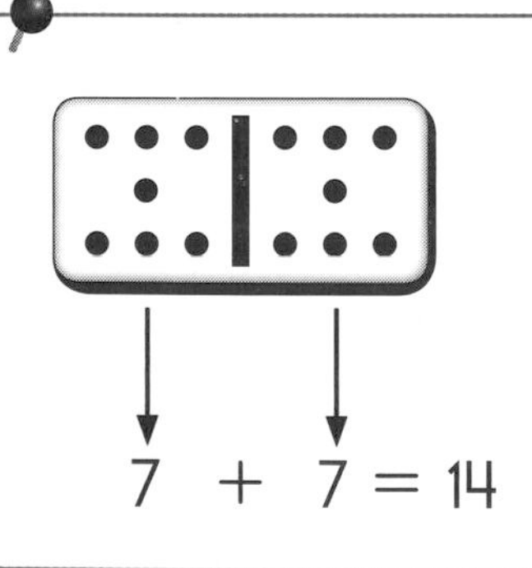

$7 + 7 = 14 \quad 14 - 7 = 7$

This **doubles fact family** uses 7, 7, and 14.

Note Doubles fact families have only two facts instead of four.

Fact triangles show the 3 numbers in a fact family.

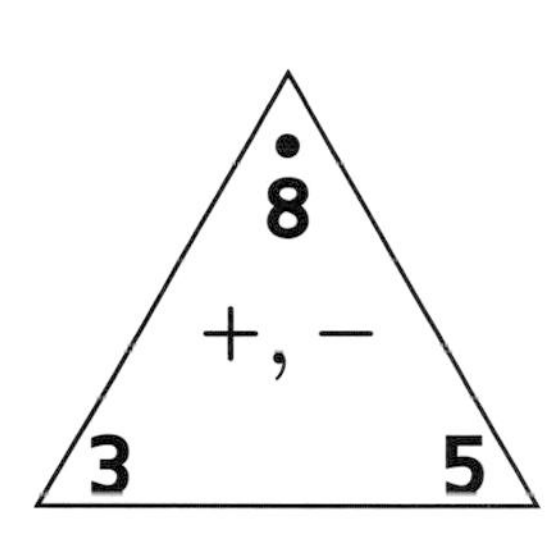

$3 + 5 = 8$ $8 - 5 = 3$

$5 + 3 = 8$ $8 - 3 = 5$

This is the fact family for 8, 5, and 3.

You can use fact triangles to practice facts.

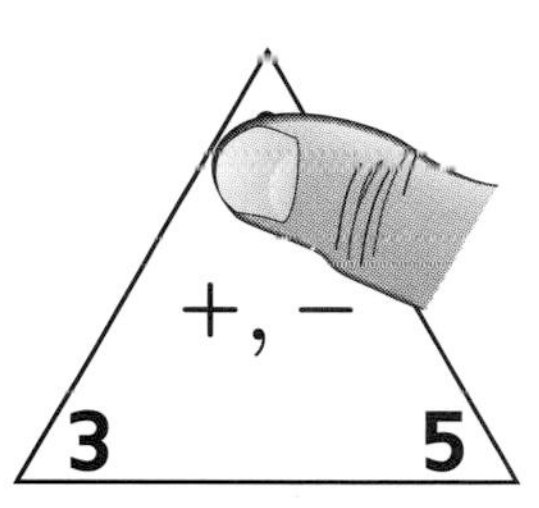

To practice addition, cover the number by the dot.

Cover 8. Think:

$3 + 5 = ?$ $5 + 3 = ?$

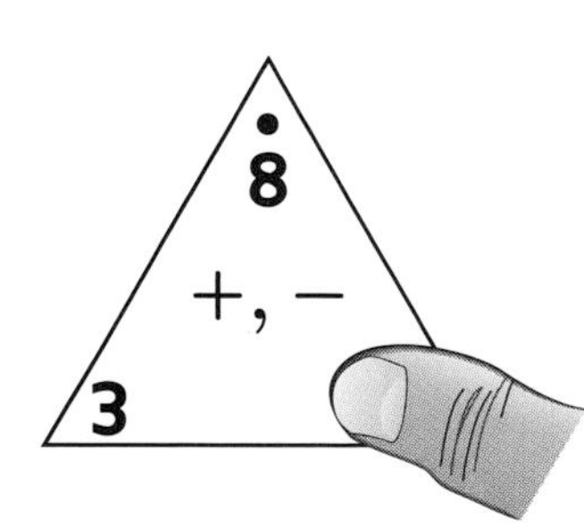

To practice subtraction, cover one of the other numbers.

Cover 5. Think:

$8 - 3 = ?$ $3 + ? = 8$

Try It Together

Use your fact triangles.
Sort out the facts you need to practice.
Think of strategies to solve those facts.

Subtraction Fact Strategies

You can **think addition** to subtract.

$9 - 5 = ?$ Think: $5 + ? = 9$.
You know $5 + \mathbf{4} = 9$, so the answer is **4.**
$9 - 5 = \mathbf{4}$

Doubles and combinations of 10 are helpful when you think addition.

$18 - 9 = ?$ Think: $9 + ? = 18$.
You know the double $9 + \mathbf{9} = 18$, so the answer is **9.**
$18 - 9 = \mathbf{9}$

$10 - 7 = ?$ Think: $7 + ? = 10$.
You know $7 + \mathbf{3}$ is a combination of 10, so the answer is **3.**
$10 - 7 = \mathbf{3}$

$13 - 6 = ?$ Think: $6 + ? = 13$.
You know the double $6 + 6 = 12$ is close, so add 1 more to get $6 + \mathbf{7} = 13$. So the answer is **7.**
$13 - 6 = \mathbf{7}$

In special cases, counting can help you quickly solve a subtraction fact.

Counting back works well when you are subtracting a small number.

$12 - 3 = ?$

Count back 3 from 12.

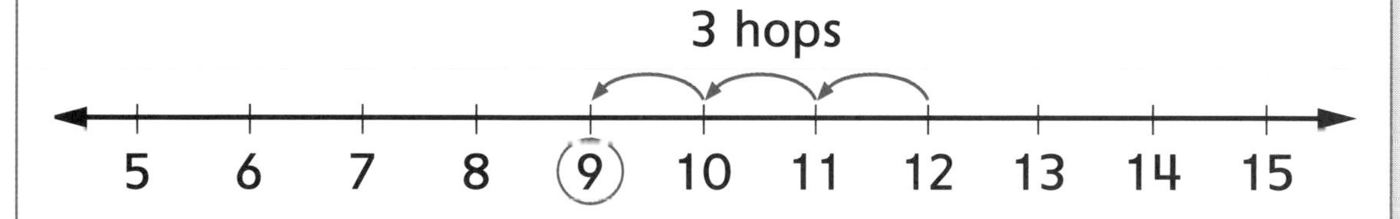

You land at **9**, so $12 - 3 = \mathbf{9}$.

Counting up works well when the numbers are close together.

$11 - 9 = ?$

Count up from 9 to 11.

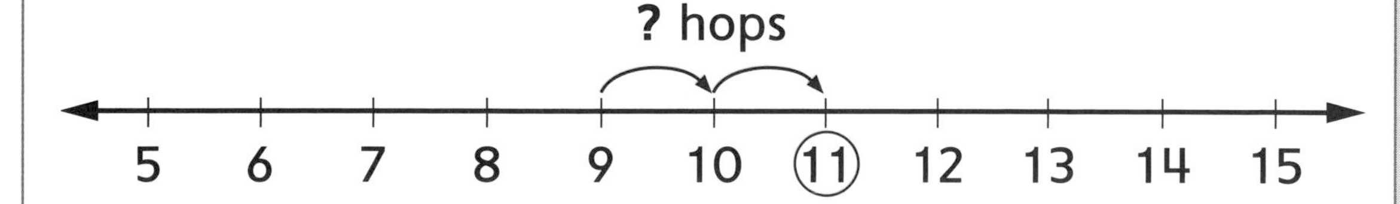

You counted up **2** hops, so $11 - 9 = \mathbf{2}$.

You can **go through 10** to solve many subtraction facts because 10 is a friendly number. You can think of going up or down through 10.

$16 - 7 = ?$

1. Think: What number should I subtract to get to 10?
2. Start by subtracting 6 to get down to 10: $16 - 6 = 10$.
3. You have only subtracted 6. You still need to subtract 1 more: $10 - 1 = \mathbf{9}$.
4. You have subtracted a total of 7 and landed at **9.** So $16 - 7 = \mathbf{9}$.

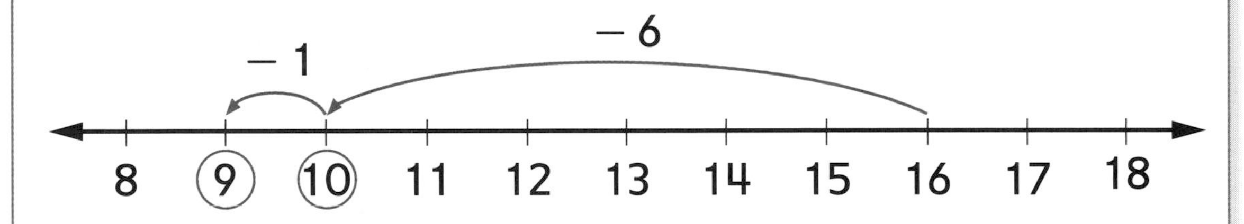

$15 - 8 = \mathbf{?}$

1. Think: $8 + \mathbf{?} = 15$.
   What number should I add to get to 10?
2. Start from 8 and count up 2 to get to 10.
3. You still have to count up 5 more to get to 15.
4. You have counted up a total of $2 + 5 = \mathbf{7}$.
   So $15 - 8 = \mathbf{7}$.

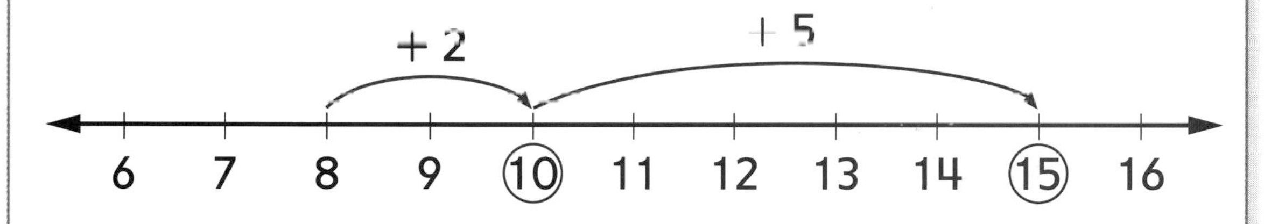

Try It Together

Sort through your fact triangles. Think about which strategies you would use for each subtraction fact.

Equal Sign

Read It Together

An **equal sign (=)** is a symbol that means "is the same amount as" or "means the same as."

$8 = 6 + 2$
8 is the same amount as 6 plus 2.

$12 - 5 = 7$
12 minus 5 means the same as 7.

A number sentence can have one number or many numbers on both sides of the equal sign.

$15 - 5 = 2 + 5 + 3$

$15 - 5$ is 10 $\quad 2 + 5 + 3$ is 10

$10 = 10$

$15 - 5$ is the same amount as $2 + 5 + 3$.

Try It Together

Are these number sentences true or false?

$5 + 2 = 8 - 1$ $\qquad 16 = 10 + 4 + 1$ $\qquad 7 = 7$

Name-Collection Box

Read It Together

A **name-collection box** is a place to write different names for the same number.

This tag names the box.

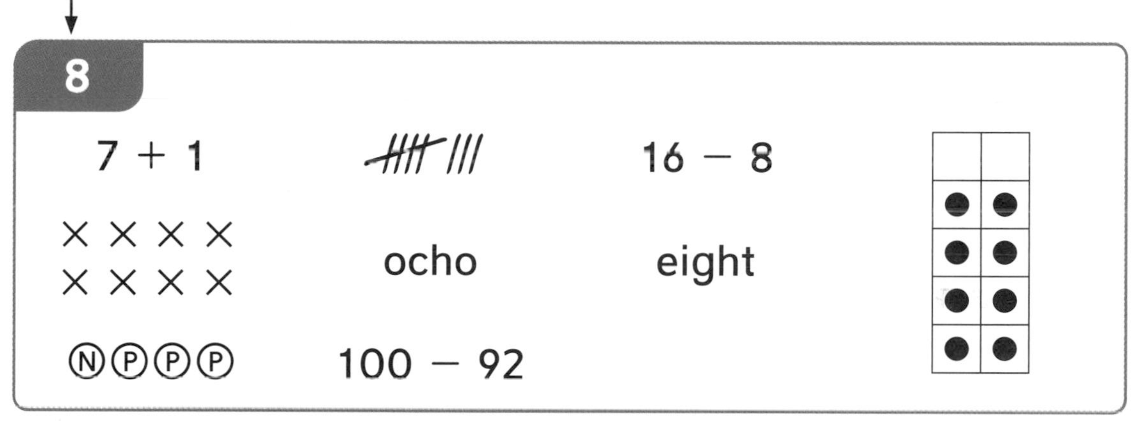

All of the names in this name-collection box are equal to 8. You can use a number sentence to check whether a name can be in the box.

$10 - 2 = 8$
This is a true number sentence.
$10 - 2$ can be in the name-collection box for 8.

Frames and Arrows

Read It Together

In a **Frames-and-Arrows diagram,** the **frames** are the shapes that hold the numbers, and the **arrows** show the path from one frame to the next.

Each diagram has a **rule box.** The **rule** in the box tells how to get from one frame to the next.

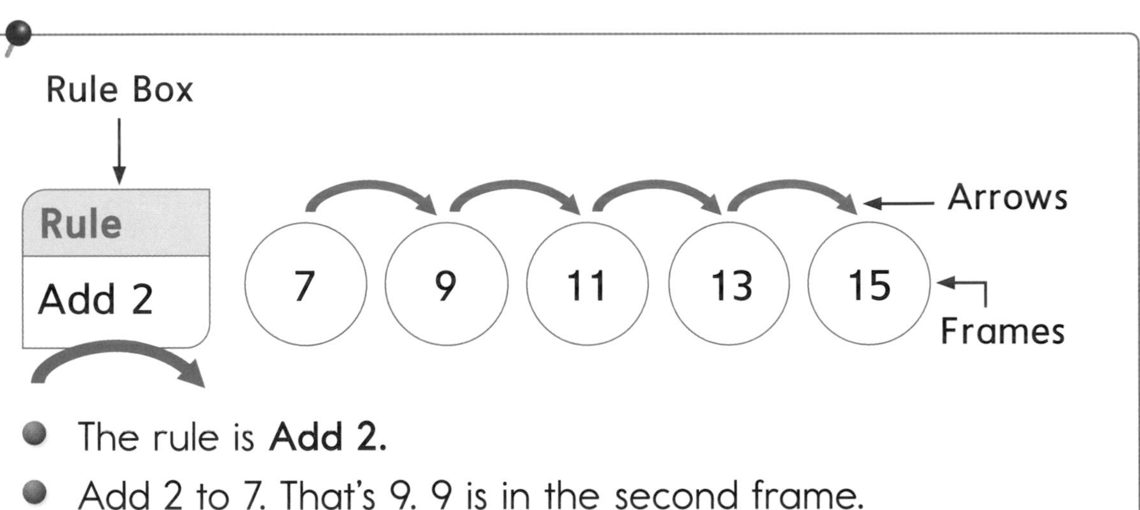

- The rule is **Add 2.**
- Add 2 to 7. That's 9. 9 is in the second frame.
  Add 2 to 9. That's 11. 11 is in the third frame.

Why is 15 in the last frame?

Use the rule to find the missing number.

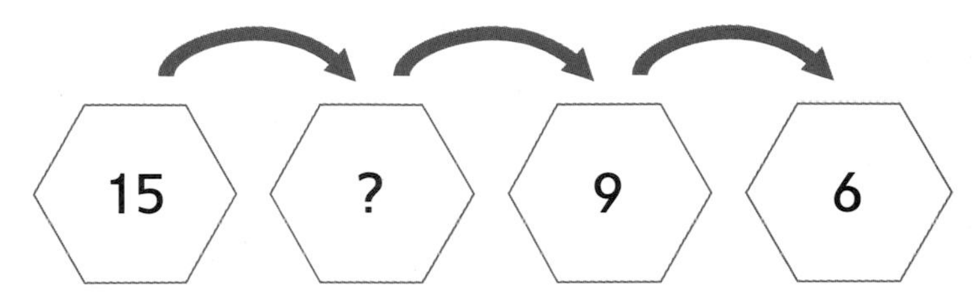

- The rule is **Subtract 3.**
  15 − 3 = **12**
  12 is the missing number.

Use the numbers in the frames to find a rule that fits.

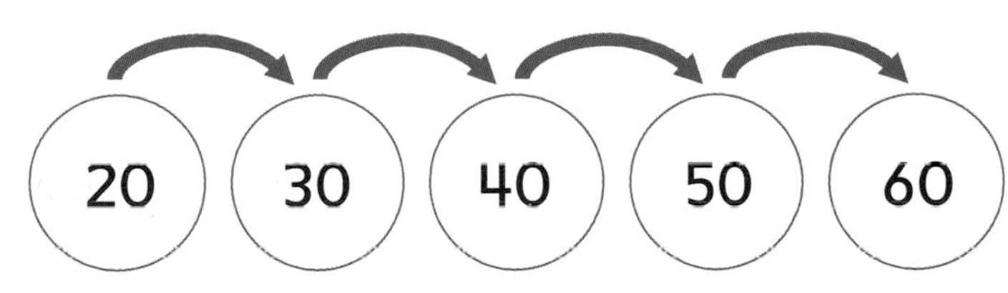

- You can use number sentences to help you find a rule.
  20 + ☐ = 30
  30 + ☐ = 40
- Each number is 10 more than the number before it.
  A rule that fits is **Add 10.**

Note Another name for this rule is **Count up 10.**

Function Machines

Read It Together

A **function machine** uses a rule to change numbers. You put a number into the machine. The machine uses the rule to change the number. The changed number comes out of the machine.

- The rule for this machine is **Add 10.**
- If you put 2 into the machine,
  it will **add 10** to 2.
  The number 12 will come out.
- If you put 4 into the machine,
  it will **add 10** to 4.
  The number 14 will come out.
- If you put 0 in the machine,
  it will **add 10** to 0.
  The number 10 will come out.

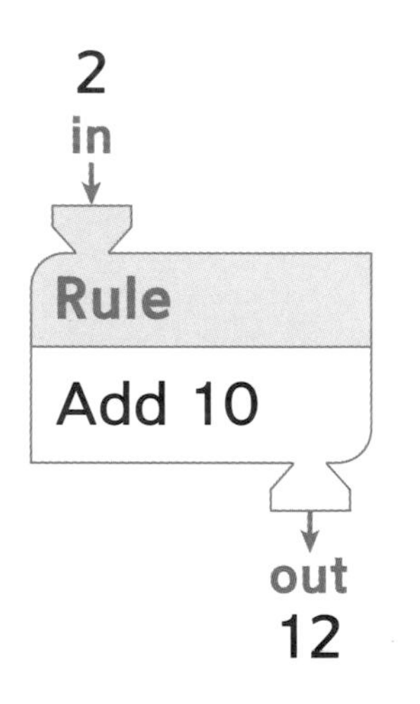

An **In and Out** table keeps track of how a function machine changes numbers.

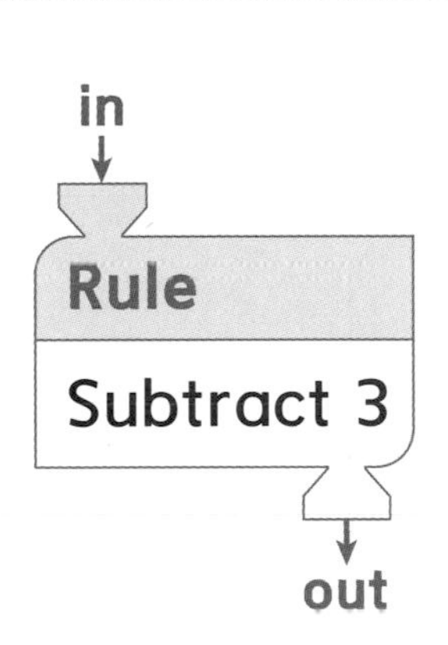

in	out
4	1
5	2
6	3
7	4

Write the numbers that are put into the machine in the **in** column.

Write the numbers that come out of the machine in the **out** column.

- If you put 4 into the machine, the machine subtracts 3 from 4.
  $4 - 3 = 1$
  1 comes out.
- If 2 comes out, then 5 was put in.
  The machine subtracted 3 from 5.
  $5 - 3 = 2$.

Use the in and out numbers to find a rule for this function machine.

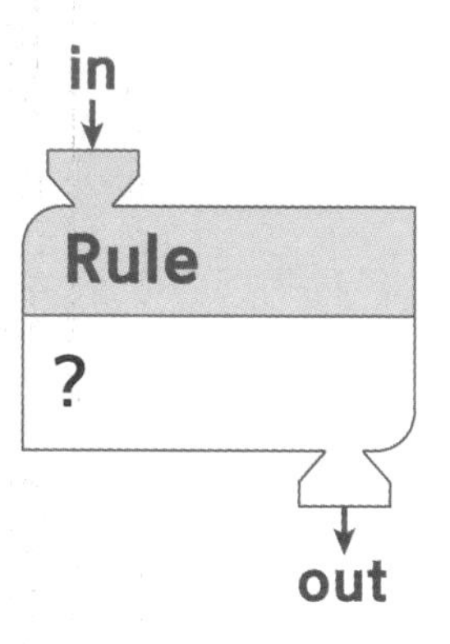

in	out
2	0
5	3
6	4
9	7

- How can you get from 2 to 0? **Subtract 2.**
  How can you get from 5 to 3? **Subtract 2.**
  Can you **subtract 2** to get from 6 to 4? Yes.
- A rule that fits is **Subtract 2.**

Try It Together

Ask your partner to write a rule. Write an In and Out table for the rule.

Patterns

Read It Together

Shapes can make **patterns.** You can tell what comes next in a pattern if you know the rule.

Some patterns repeat over and over.

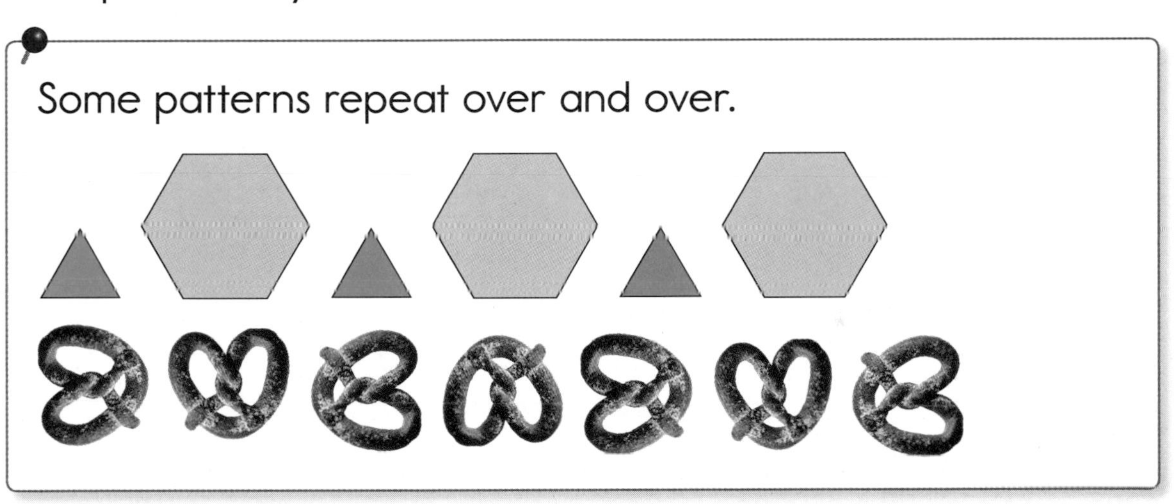

Numbers can also make patterns.

Even numbers and odd numbers can make dot patterns.

Even Numbers

2 4 6 8

Odd Numbers

1 3 5 7

A number grid has many patterns.

This number grid shows odd and even numbers.
The odd numbers are green.
The even numbers are orange.

−9	−8	−7	−6	−5	−4	−3	−2	−1	(0)
1	2	3	4	5	6	7	8	9	(10)
11	12	13	14	15	16	17	18	19	(20)
21	22	23	24	25	26	27	28	29	(30)
31	32	33	34	35	36	37	38	39	(40)
41	42	43	44	45	46	47	48	49	(50)
51	52	53	54	55	56	57	58	59	(60)
61	62	63	64	65	66	67	68	69	(70)
71	72	73	74	75	76	77	78	79	(80)
81	82	83	84	85	86	87	88	89	(90)
91	92	93	94	95	96	97	98	99	(100)
101	102	103	104	105	106	107	108	109	(110)

The circled numbers show counting by 10s starting at 0.
You can see a pattern.
All the circled numbers end in 0.

Numbers and Operations in Base Ten

Numbers All Around

Read It Together

Numbers are used in many ways.

Count the shells. How many shells are there?

Numbers are used for **counting.**

Numbers are used as **measures.**

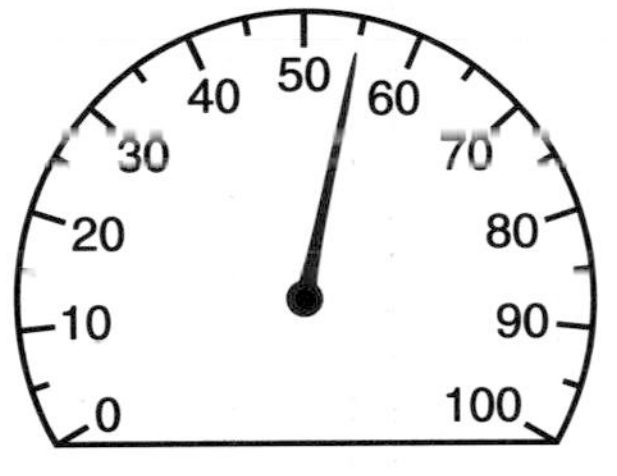

55 miles per hour

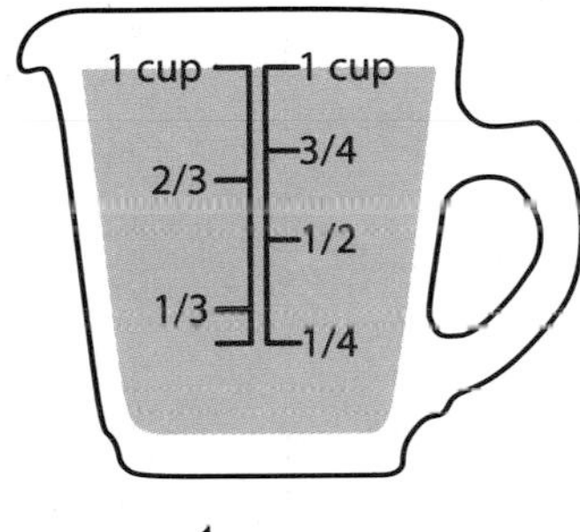

1 cup

50 pounds

Numbers are used to **compare.**

A dime is worth 5 cents more than a nickel.

There are 3 more boys than girls.

Numbers are used as **codes.**

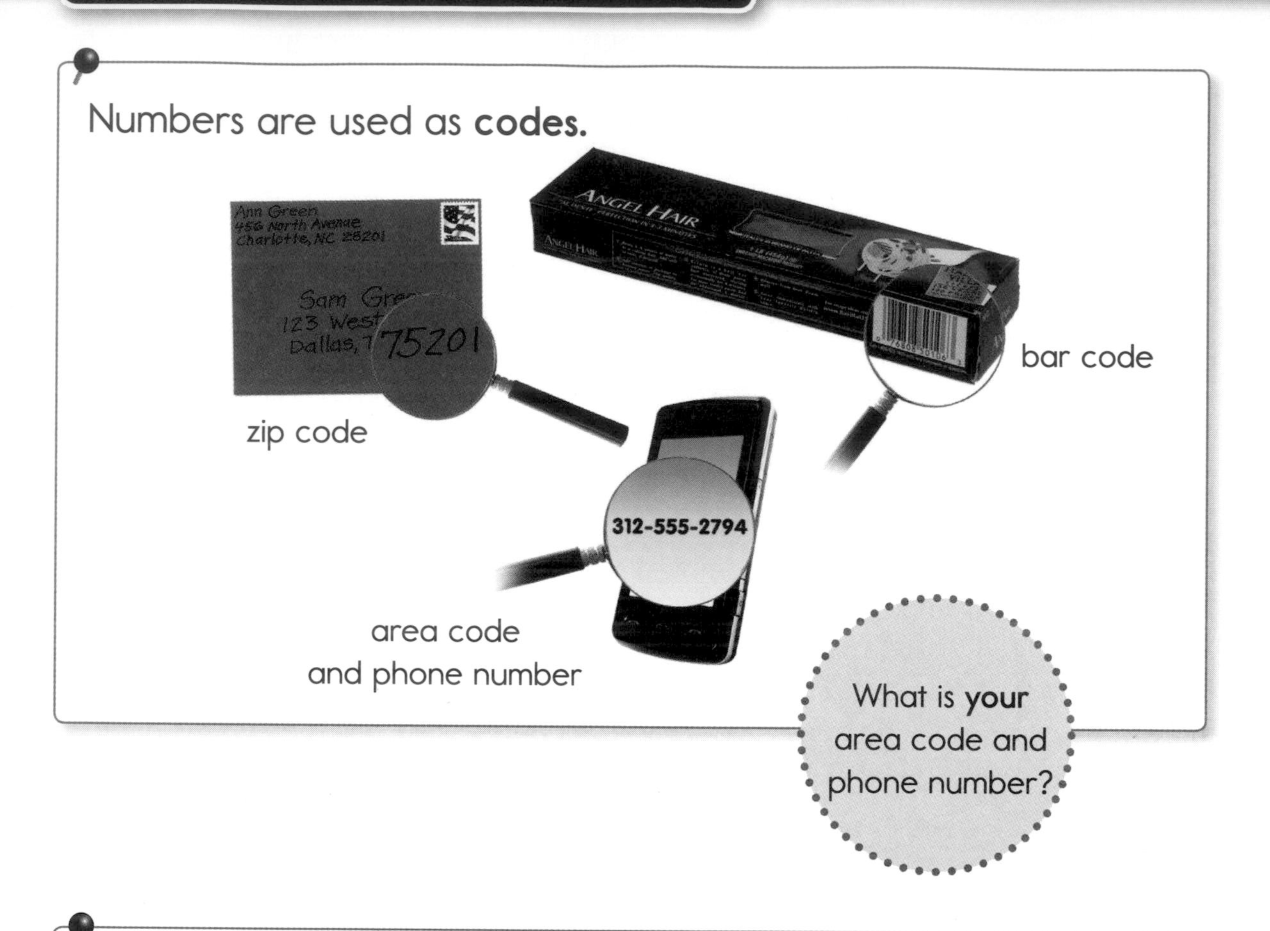

Numbers are used to **show locations.**

We live at
35 Park Street.

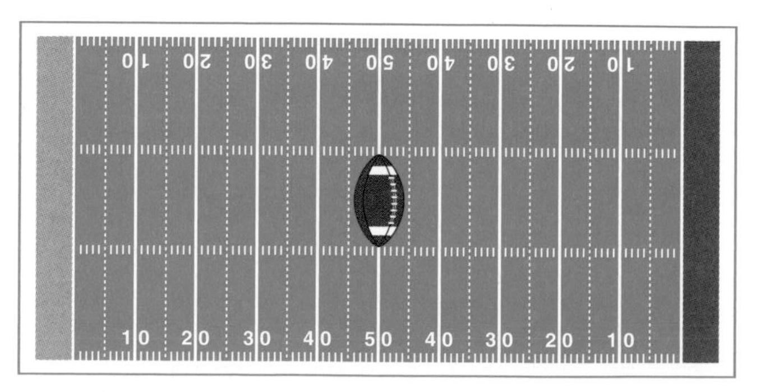

The football is on the 50-yard line.

Numbers are used for **ordering.**

Note When you use numbers, say what you are counting or measuring. That is called a **unit**.

5 **pennies**
50 **pounds**
5th **child**

unit

Try It Together

Look around the room. Try to find numbers used in different ways.

Counting Tools

Read It Together

Number lines and **number grids** are tools for counting. To count on a **number line,** think about hopping from one number to another.

Count by 5s. Start at 0.

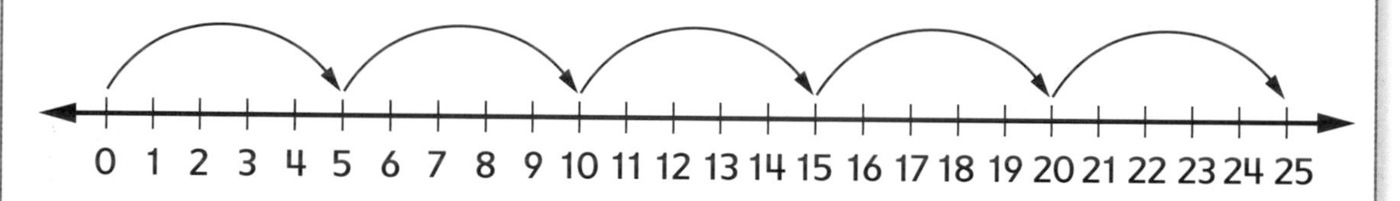

Say: 0, 5, 10, 15, 20, 25

Count back by 1s. Start at 10.

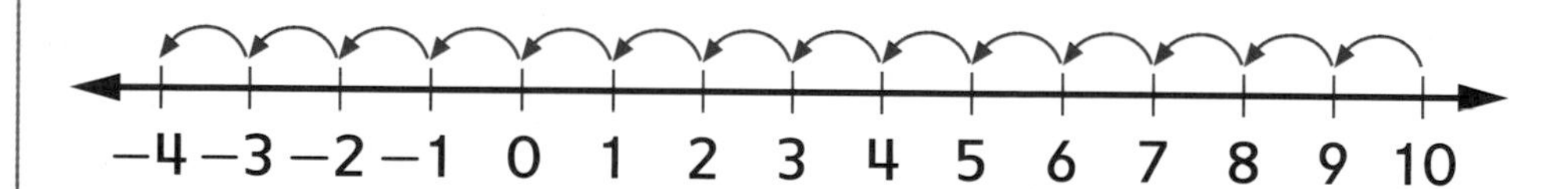

Say: 10, 9, 8, 7, 6, 5, 4, 3, 2, 1, 0, −1, −2, −3, −4

Count by 2s. Start at 0.

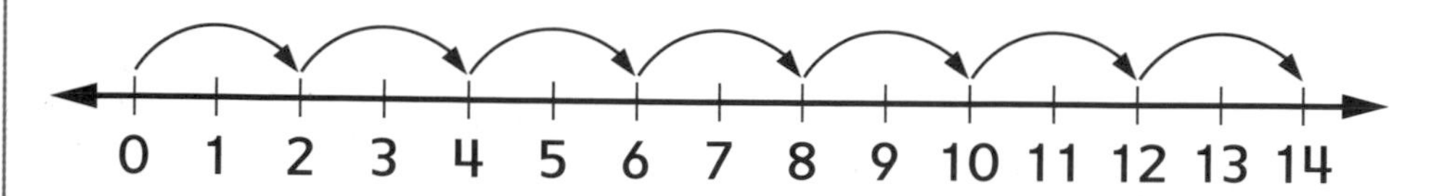

Say: 0, 2, 4, 6, 8, 10, 12, 14

Numbers on a **number grid** are in rows and columns.

−9	−8	−7	−6	−5	−4	−3	−2	−1	0
1	2	3	4	5	6	7	8	9	10
11	12	13	14	15	16	17	18	19	20
21	22	23	24	25	26	27	28	29	30
31	32	33	34	35	36	37	38	39	40
[illegible]	42	43	44	45	46	47	48	49	50
51	52	53	54	55	56	57	58	59	60
61	62	63	64	65	66	67	68	69	70
71	72	73	74	75	76	77	78	79	80
[illegible]	82	83	84	85	86	87	88	89	90
91	92	93	94	95	96	97	98	99	100
101	102	103	104	105	106	107	108	109	110

Note To move from row to row, follow the matching colors.

10 → 11

−9	−8	−7	−6	−5	−4	−3	−2	−1	0
1	2	3	4	5	6	7	8	9	10
11	12	13	14	15	16	17	18	19	20
21	22	23	24	25	26	27	28	29	30
31	32	33	34	35	36	37	38	39	40
41	42	43	44	45	46	47	48	49	50
51	52	53	54	55	56	57	58	59	60
61	62	63	64	65	66	67	68	69	70
71	72	73	74	75	76	77	78	79	80
81	82	83	84	85	86	87	88	89	90
91	92	93	94	95	96	97	98	99	100
101	102	103	104	105	106	107	108	109	110

When you move to the right, numbers get *larger* by 1.
For example: 15 is 1 *more* than 14.

When you move to the left, numbers get *smaller* by 1.
For example: 23 is 1 *less* than 24.

When you move down, numbers get *larger* by 10.
For example: 37 is 10 *more* than 27.

When you move up, numbers get *smaller* by 10.
For example: 43 is 10 *less* than 53.

Estimation

Read It Together

An **estimate** is an answer that is close to an exact answer. You make estimates every day. When you say the word *about* before a number, that number is an estimate.

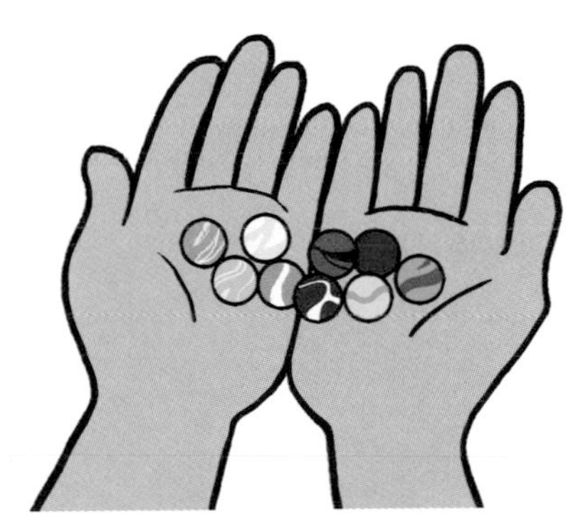

This is about 10 marbles.

If you do not need to know exactly how many objects there are, you can make an estimate. One way to estimate a large number of objects is to look at a smaller part.

About how many cubes are in the jar?

There are about 25 cubes on the top layer and 6 layers of cubes in the jar. Four 25s is 100 and two 25s is 50. 100 plus 50 equals 150.

There are about 150 cubes in the jar.

Place Value

Read It Together

Our number system is a base-10 system.

One ten is a bundle of 10 ones.

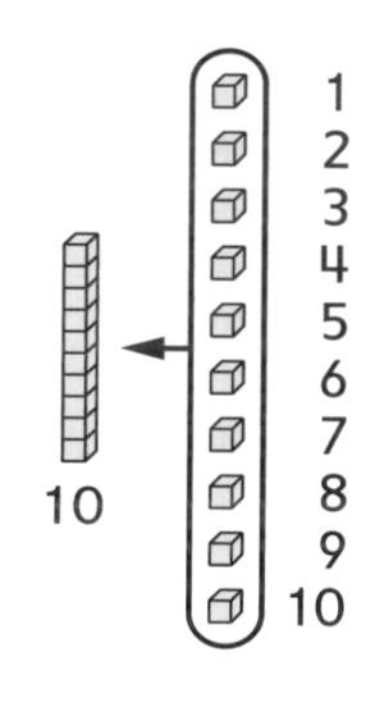

A **teen number** is made up of a ten and some ones.

10 11 12 13 14 15

The blocks show fifteen.

One hundred is a bundle of 10 tens.

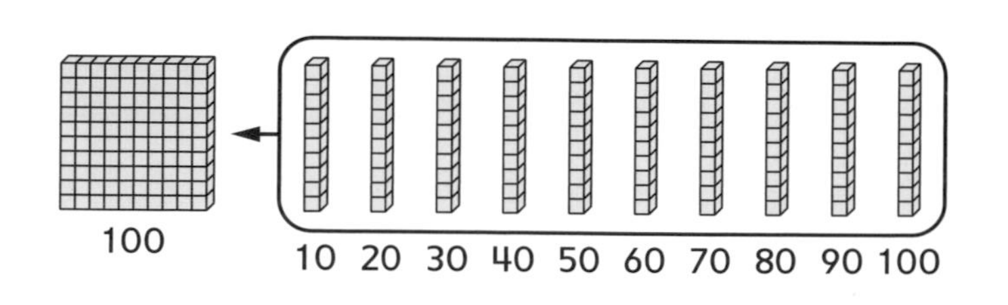

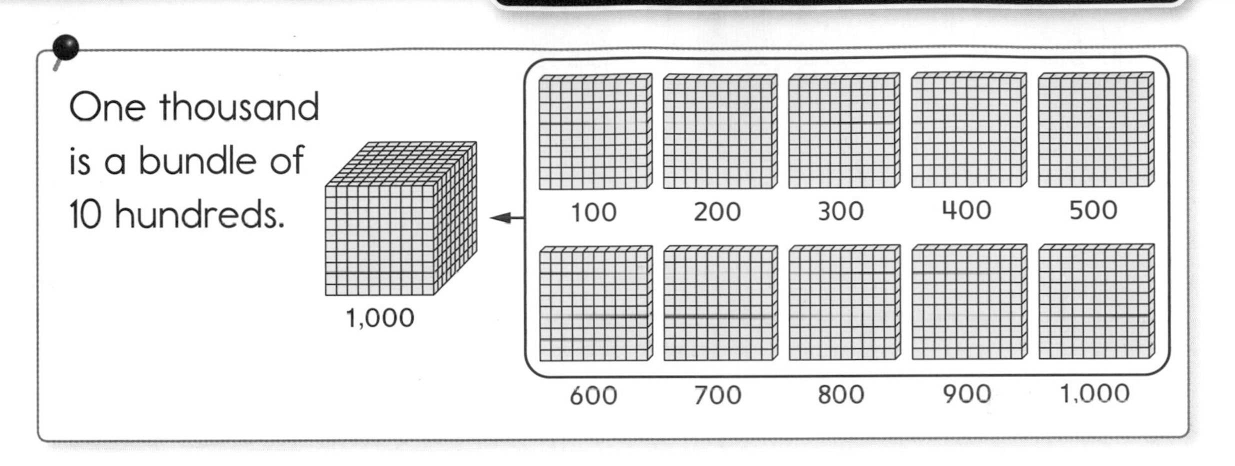

Base-10 blocks can show numbers.

Base-10 block	Base-10 shorthand	Name	Value
		cube	1
		long	10
		flat	100
		big cube	1,000

You can show numbers in different ways.

You can show 43 in different ways.

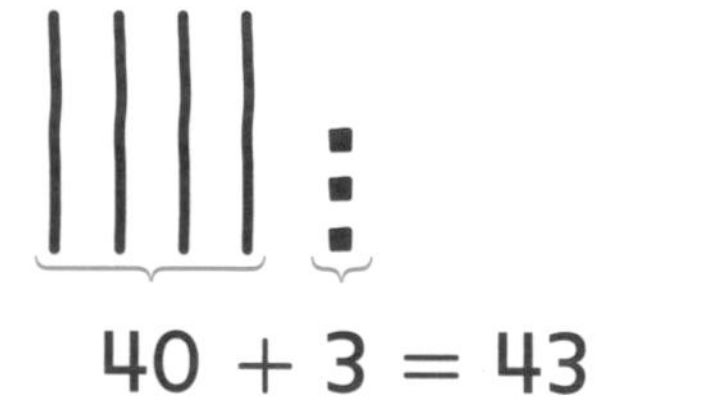

40 + 3 = 43   30 + 13 = 43

You can show 235 in different ways.

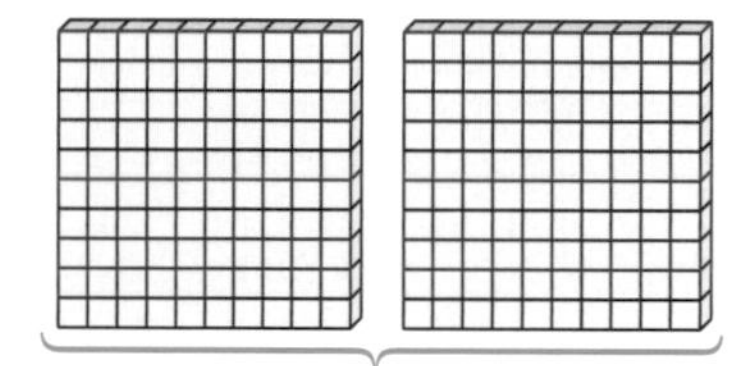 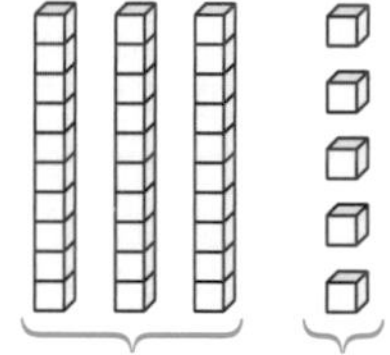

200 + 30 + 5 = 235

100 + 130 + 5 = 235

Find another way to show 235 with base-10 blocks.

Numbers can be written using these **digits.**

0	1	2	3	4	5	6	7	8	9

The place of a digit in a number tells how much the digit is worth.

tens	ones
5	0

The **5** in the **tens** place is worth **50.**
There are **0 ones.**
The number shown is 50.
Say: fifty

hundreds	tens	ones
3	4	2

The **3** in the **hundreds** place is worth **300.**
The **4** in the **tens** place is worth **40.**
The **2** in the **ones** place is worth **2.**
The number shown is 342.
Say: three hundred forty-two

Comparing Numbers

Read It Together

You can use place value to compare numbers.

Which is larger, 240 or 204?

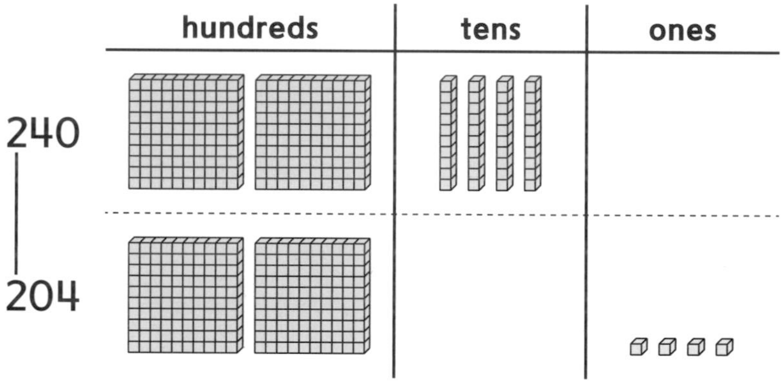

	hundreds	tens	ones
240			
204			

The 2 represents 2 hundreds in each number.
In 240, the 4 represents 4 tens.
In 204, the 0 represents 0 tens.
There are more tens in 240 than 204.
So 240 is greater than 204.

You can use symbols to **compare** numbers.

$25 > 20$
25 **is greater than** 20.

$20 < 25$
20 **is less than** 25.

$20 = 20$
20 **is the same as** 20.
20 **is equal to** 20.

The symbol $\neq$ means **is not equal to.**

To help you remember what $>$ and $<$ mean, think about an alligator.

The alligator eats the larger number.

$31 > 18$
31 **is greater than** 18.

$205 < 250$
205 **is less than** 250.

Adding Larger Numbers

Read It Together

There are many different ways to add larger numbers.

Try this ⟶ 17 + 25 = **?**

You can use base-10 blocks to add.

1. Show 17 with 1 long and 7 cubes.

2. Show 25 more with 2 longs and 5 cubes.
3. Count all the blocks.

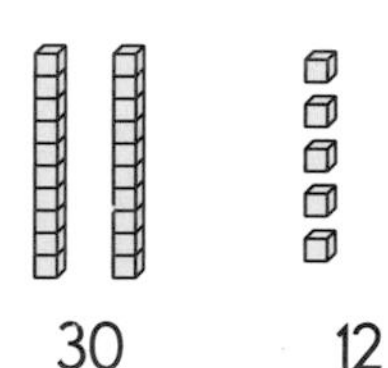

17 + 25 = **42**

You can trade blocks to make counting easier.

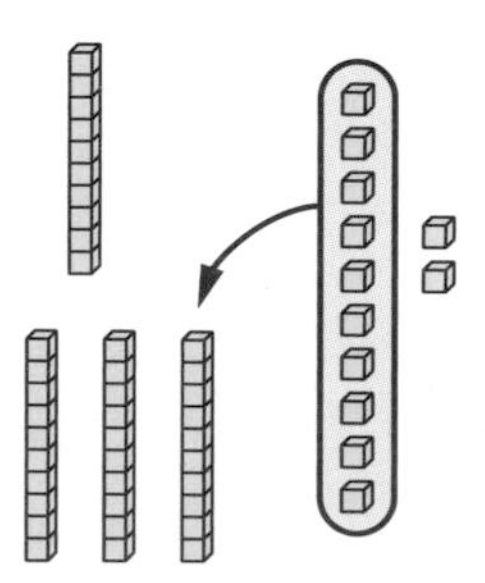

Trade 10 cubes for 1 long.

17 + 25 = **42**

You can use a **number grid** to add.

−9	−8	−7	−6	−5	−4	−3	−2	−1	0
1	2	3	4	5	6	7	8	9	10
11	12	13	14	15	16	17	18	19	20
21	22	23	24	25	26	27	28	29	30
31	32	33	34	35	36	37	38	39	40
41	42	43	44	45	46	47	48	49	50
51	52	53	54	55	56	57	58	59	60
61	62	63	64	65	66	67	68	69	70
71	72	73	74	75	76	77	78	79	80
81	82	83	84	85	86	87	88	89	90
91	92	93	94	95	96	97	98	99	100
101	102	103	104	105	106	107	108	109	110

Note At the end of a row, go to the beginning of the next row and keep counting.

17 + 25 = **?**

1. Start at 17.
2. Add 20.
   - Add 2 tens by moving down 2 rows to 37.
3. Add 5.
   - Add 5 ones by counting on 5 spaces.

17 + 25 = **42**

You can use an open number line to add.

$17 + 25 = ?$

1. Draw a line. Mark and label point 17.
2. Think: 25 = 2 tens and 5 ones.
3. Start at 17. **Count up** 2 tens. Mark points at 27 and 37. **Count up** 5 ones from 37. Mark each point.

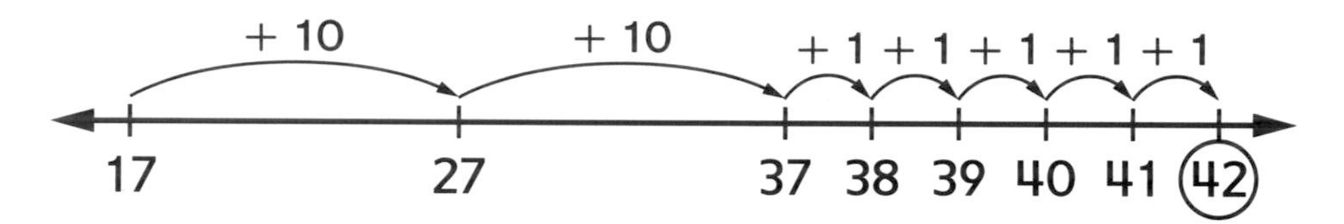

You end at 42. 17 + 25 = **42**

$398 + 132 = ?$

1. Draw a line. Mark and label point 398.
2. Think: 132 = 100 + 30 + 2. Choose what to add first to make counting up easy.
3. Start at 398. **Count up** 2. **Count up** 100. **Count up** 30. Mark points at 400, 500, and 530.

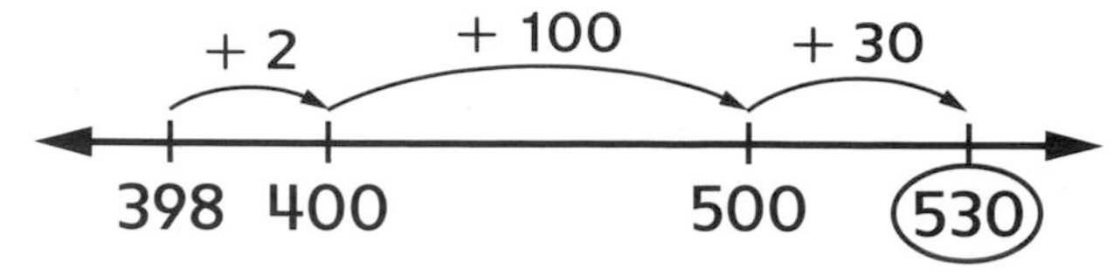

You end at 530. 398 + 132 = **530**

Sometimes estimates are called **ballpark estimates.** You can make ballpark estimates by finding **close-but-easier** numbers.

You can use ballpark estimates to solve problems when an exact answer is not needed.

There are 100 cartons of milk in the lunchroom.
53 first graders need milk.
38 second graders need milk.
Is there enough milk in the lunchroom for the first and second graders?

$$53 + 38 = ?$$
$$\downarrow \quad \downarrow$$
$$50 + 40 = 90$$

close-but-easier numbers

About 90 cartons of milk are needed.
So 100 cartons is enough.

You can use a ballpark estimate to see if an exact answer is reasonable.

$$135 + 387 = ?$$
$$\downarrow \quad \downarrow$$

close-but-easier numbers $100 + 400 = 500$

A sum close to 500 makes sense for 135 + 387.

You can use **partial-sums addition** to add.

$$135 + 387 = ?$$

Think: $135 = 100 + 30 + 5$
$387 = 300 + 80 + 7$

$$\begin{array}{r} 135 \\ +\ 387 \\ \hline \end{array}$$

1. Add the 100s. $100 + 300 =$ 400
2. Add the 10s. $30 + 80 =$ 110
3. Add the 1s. $5 + 7 =$ 12
4. Add the partial sums. 522

$$135 + 387 = \mathbf{522}$$

Look at the estimate on page 79.

522 is close to 500, so 522 makes sense.

Note If you can add the 100s, 10s, and 1s in your head, then you don't need to write the numbers in green.

Subtracting Larger Numbers

Read It Together

There are many different ways to subtract larger numbers.

Try this ⟶ 43 − 27 = **?**

You can use **base-10 blocks** to subtract.

1. Show 43 with 4 longs and 3 cubes.

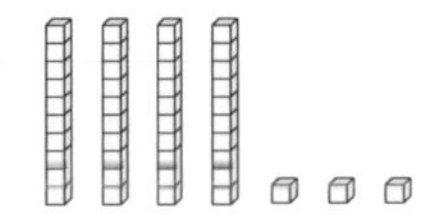

2. Take away 2 longs or 20.

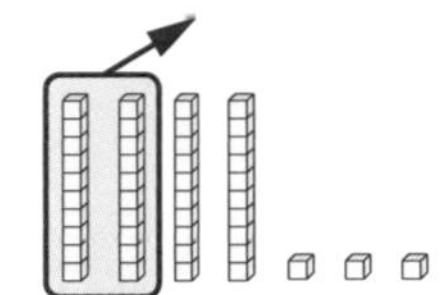

3. Cannot take away 7 cubes.
   Trade 1 long for 10 cubes.

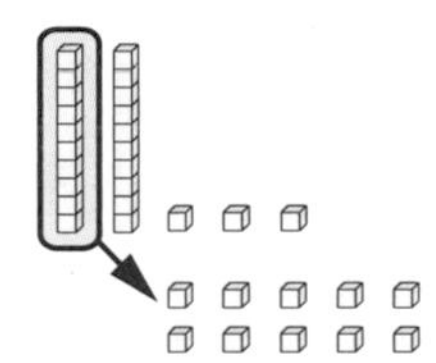

4. Take away 7 cubes or 7.

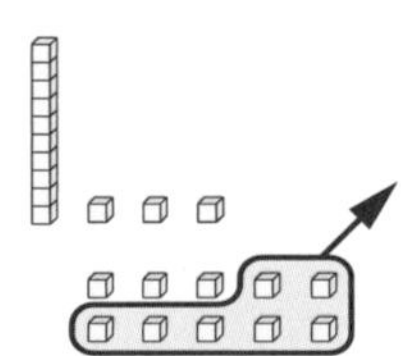

5. 1 long and 6 cubes are left.
   So 1 ten and 6 ones or 16 is left.

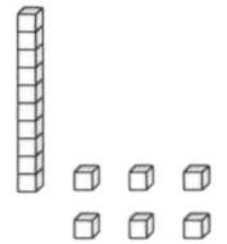

43 − 27 = **16**

You can use a **number grid** to subtract.

−9	−8	−7	−6	−5	−4	−3	−2	−1	0
1	2	3	4	5	6	7	8	9	10
11	12	13	14	15	16	17	18	19	20
21	22	23	24	25	26	27	28	29	30
31	32	33	34	35	36	37	38	39	40
41	42	43	44	45	46	47	48	49	50
51	52	53	54	55	56	57	58	59	60
61	62	63	64	65	66	67	68	69	70
71	72	73	74	75	76	77	78	79	80
81	82	83	84	85	86	87	88	89	90
91	92	93	94	95	96	97	98	99	100
101	102	103	104	105	106	107	108	109	110

$43 - 27 = ?$

1. Start at 43.
2. Subtract 20.
   - Subtract 2 tens by moving up 2 rows to 23.
3. Subtract 7.
   - Subtract 7 ones by counting back 7 spaces to 16.

$43 - 27 = \mathbf{16}$

Another way to subtract on a number grid is to count up.

−9	−8	−7	−6	−5	−4	−3	−2	−1	0
1	2	3	4	5	6	7	8	9	10
11	12	13	14	15	16	17	18	19	20
21	22	23	24	25	26	27	28	29	30
31	32	33	34	35	36	37	38	39	40
41	42	43	44	45	46	47	48	49	50
51	52	53	54	55	56	57	58	59	60
61	62	63	64	65	66	67	68	69	70
71	72	73	74	75	76	77	78	79	80
81	82	83	84	85	86	87	88	89	90
91	92	93	94	95	96	97	98	99	100
101	102	103	104	105	106	107	108	109	110

$43 - 27 = ?$

1. Start at 27.
2. Count up 10.
   - Count up 1 ten by moving down 1 row to 37.
3. Count up 6 more.
   - Count 6 ones by counting up 6 spaces to 43.
4. You counted up 1 ten and 6 ones. You counted up 16.

$43 - 27 = \mathbf{16}$

You can use an open number line to subtract.

$$52 - 34 = ?$$

1. Draw a line. Mark and label point 52.

52

2. Think: 34 = 3 tens and 4 ones.
3. Start at 52. **Count back** 3 tens.
   - Mark points at 42, 32, and 22.
   - **Count back** 4 ones from 22. Mark each point.

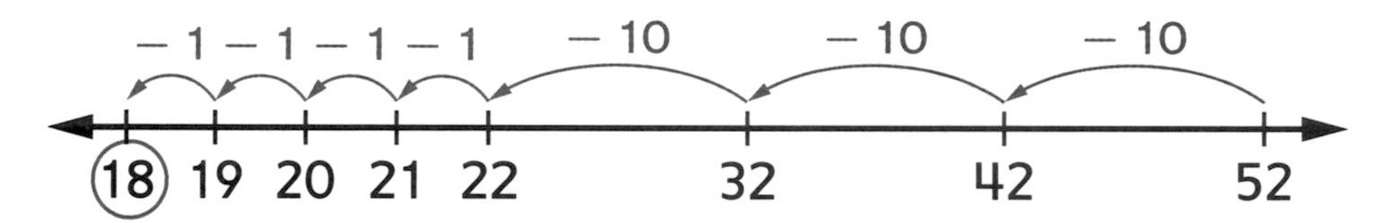

You end at 18.

$$52 - 34 = \mathbf{18}$$

You can count up on an open number line to subtract.

$$52 - 34 = ?$$

1. Draw a line. Mark and label point 34.

2. Think: How can I get from 34 to 52?

3. Start at 34. **Count up** by tens and then by ones.

- **Count up** 1 ten. Mark 44.
- **Count up** by ones from 44. Mark each point.

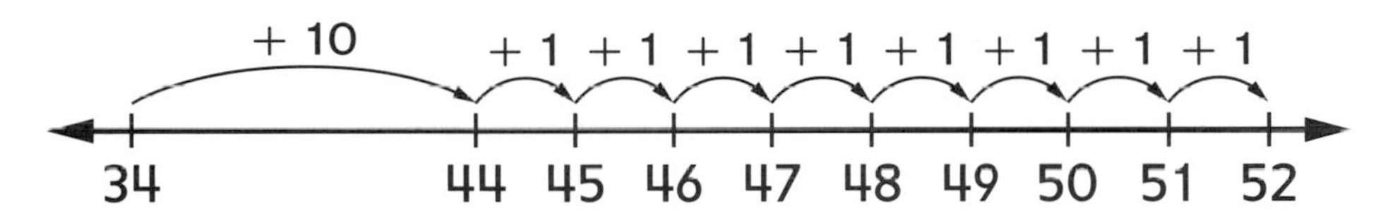

4. You counted up 1 ten and 8 ones. You counted up **18**.

$$52 - 34 = \mathbf{18}$$

Try It Together

Subtract two larger numbers. Show a partner what you did.

You can make ballpark estimates for differences by finding **close-but-easier** numbers.

You can use ballpark estimates to solve problems when an exact answer is not needed.

Darlene won 73 tickets at the school carnival.
She gave 27 to her sister.
Darlene needs 35 tickets to buy a pen. Can she buy it?

$$73 - 27 = ?$$

close-but-easier numbers $70 - 30 = 40$

Darlene has about 40 tickets left. She can buy the pen.

You can use a ballpark estimate to see if an exact answer is reasonable.

$$352 - 168 = ?$$

close-but-easier numbers $350 - 200 = 150$

A difference close to 150 would make sense for the answer to $352 - 168$.

You can use **expand-and-trade subtraction** to subtract.

$43 - 27 = ?$

1. Think of each number in expanded form.

   $43 = 40 + 3$
   $27 = 20 + 7$

2. Use the expanded form to help you write the problem so you can see the tens and ones.

$$\begin{array}{r} 43 \rightarrow 40 + 3 \\ -\ 27 \rightarrow \underline{20 + 7} \end{array}$$

3. Look at the tens. Since $40 > 20$, you do not need to make a trade.

4. Look at the ones. Since $3 < 7$, you need to make a trade. So, trade 1 ten for 10 ones.

$$\begin{array}{r} \ 30 \quad 13 \\ 43 \rightarrow \cancel{40} + \cancel{3} \\ -\ 27 \rightarrow \underline{20 + 7} \end{array}$$

Think: Is $30 + 13$ still 43 altogether?

Yes, so you have made a good trade.

5. Subtract the tens and the ones. $10 + 6 = 16$, so 16 is the answer.

$$\begin{array}{r} \ 30 \quad 13 \\ 43 \rightarrow \cancel{40} + \cancel{3} \\ -\ 27 \rightarrow \underline{20 + 7} \\ 10 + 6 = 16 \end{array}$$

$43 - 27 = \mathbf{16}$

Look at page 81. How is expand-and-trade subtraction like using base-10 blocks? How is it different?

You can use **expand-and-trade subtraction** to subtract numbers larger than two digits.

$$352 - 168 = ?$$

1. Think of each number in expanded form.

$$352 = 300 + 50 + 2$$
$$168 = 100 + 60 + 8$$

2. Use the expanded form to help you write the problem so you can see the hundreds, tens, and ones.

$$\begin{array}{rcr} 352 & \rightarrow & 300 + 50 + 2 \\ -\ 168 & \rightarrow & 100 + 60 + 8 \\ \hline \end{array}$$

3. Look at the hundreds. Since 300 > 100, you do not need to make a trade.

4. Look at the tens. Since 50 < 60, you need to make a trade. So, trade 1 hundred for 10 tens.

$$\begin{array}{rcr} & & 200 \quad\ 150 \qquad \\ 352 & \rightarrow & \cancel{300} + \cancel{50} + 2 \\ -\ 168 & \rightarrow & 100 + 60 + 8 \\ \hline \end{array}$$

Think: Is 200 + 150 + 2 still 352 altogether?
Yes, so you have made a good trade.

5 Look at the ones. Since $2 < 8$, you need to make a trade. So trade 1 ten for 10 ones.

```
                140
         200    ~~150~~   12
352 → ~~300~~ + ~~50~~ + ~~2~~
−168 → 100 + 60 + 8
```

Think: Is $200 + 140 + 12$ still 352 altogether?
Yes, so you have made a good trade.

6 Subtract the hundreds, the tens, and the ones. $100 + 80 + 4 = 184$, so 184 is the answer.

```
                140
         200    ~~150~~   12
352 → ~~300~~ + ~~50~~ + ~~2~~
−168 → 100 + 60 + 8
       100 + 80 + 4 = 184
```

$352 - 168 = \mathbf{184}$

Look at the estimate on page 86.
184 is close to 150, so 184 makes sense.

Try It Together

Write a subtraction problem with larger numbers. Solve it using expand-and-trade subtraction.

You can use addition and subtraction with larger numbers to solve number stories.

Jack rode his bike for 23 minutes on Monday.
He rode it for 31 minutes on Tuesday.
How many minutes did he ride his bike in all?

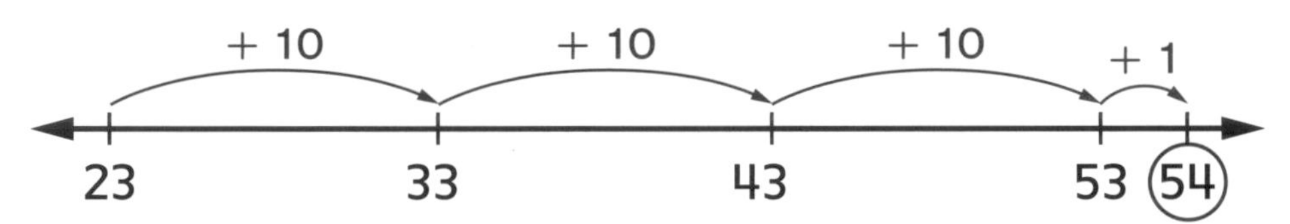

Jack rode his bike for **54** minutes.

Larissa had 50 stickers in her sticker book.
She gave 17 stickers to her friend.
How many stickers does she have left?

21	22	23	24	25	26	27	28	29	30
31	32	33	~~34~~	~~35~~	~~36~~	~~37~~	~~38~~	~~39~~	~~40~~
41	42	43	44	45	46	47	48	49	50

Larissa has **33** stickers left.

Make ballpark estimates to see whether the answers are reasonable.

Numbers All Around: Counting

Counting is an important part of our lives. People count every day.

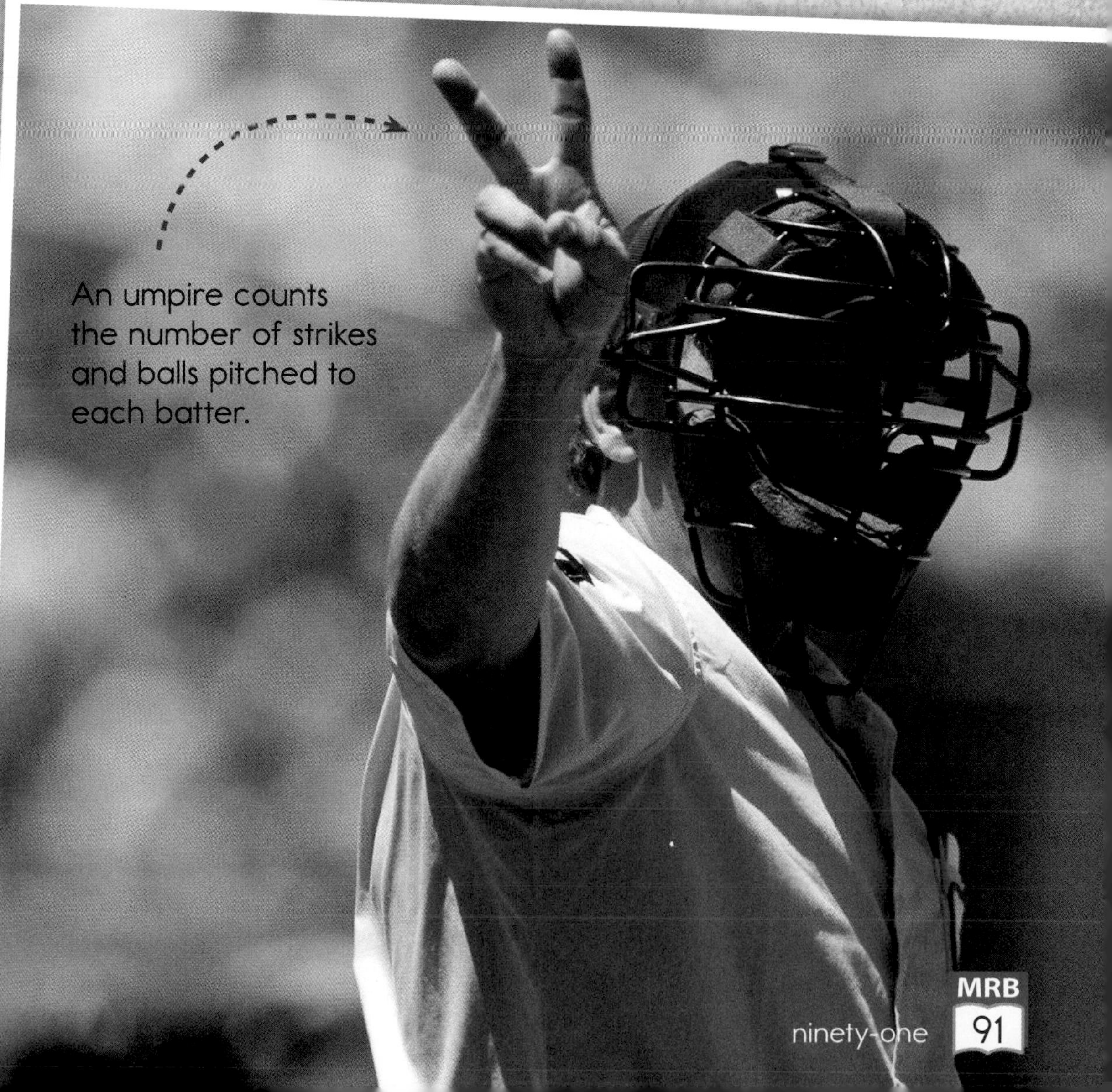

An umpire counts the number of strikes and balls pitched to each batter.

Photodisc/Getty Images

Swimmers count their strokes so they know when to turn.

Workers use machines to count the coins collected each day at banks. How might you count 10,000 pennies?

Herpetologists study reptiles and amphibians. They count the growth rings of box turtles to find out about how old they are.

Wildlife researchers have to estimate the number of birds in a large flock. Why might wildlife researchers need to count birds?

Hematologists are blood specialists. They count red blood cells to see how healthy a person is.
What things can *you* count?

Measurement and Data

Length

Read It Together

You can compare the **length** of objects. To compare, line up one end of each of the objects.

You can compare the distance from end to end.

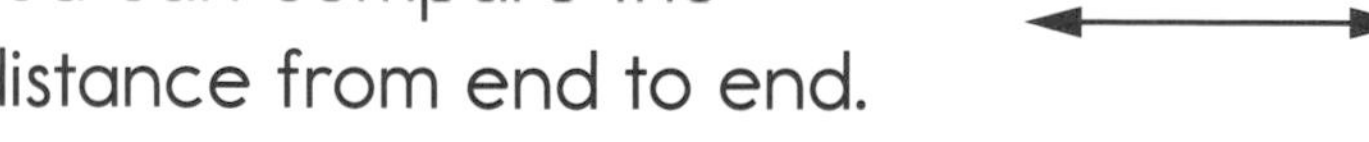

- The baby spoon is the shortest.
- The measuring spoon is longer than the baby spoon.
- The dinner spoon is the longest.

You can compare the distance across each spoon.

- The measuring spoon is the widest.
- The dinner spoon is wider than the baby spoon.

You can use an object to compare lengths.

The distance from the bottom of the book to the top of the book is *greater* than the length of the spoon.

The distance across the book is a little *less* than the length of the spoon.

So, the distance from bottom to top is greater than the distance across the book.

You can measure an object using other objects as **units.**

Emma and her sister, Ava, measured a dinner spoon. They both used paper clips as units, but their measurements are not the same.

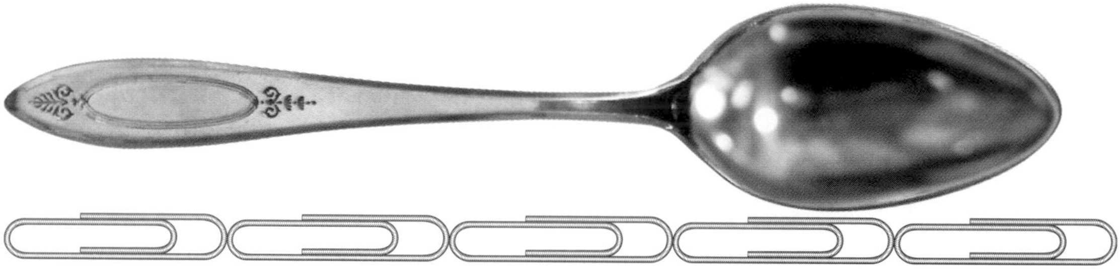

Emma said, "I think the spoon is about 5 paper clips long."

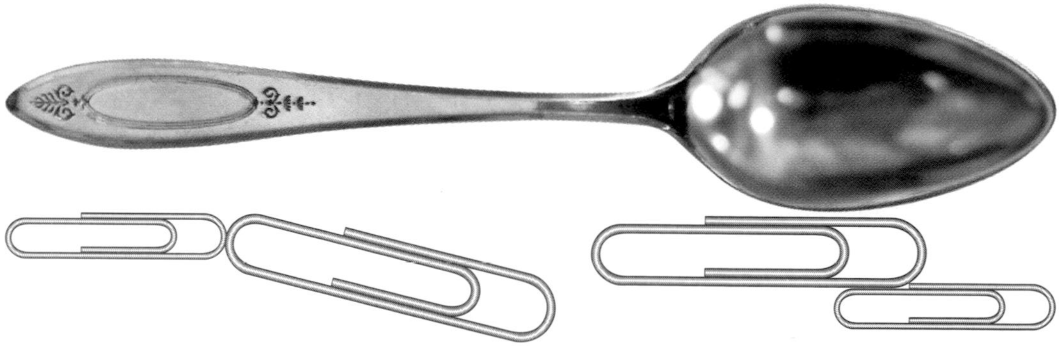

Ava said, "I think the spoon is about 4."

Which measurement do you agree with? Why?

Emma's measurement is better than Ava's. Here is what to do to make good measurements like Emma's:

- Use all the same-size units. For example, use all small paper clips. Do not use some large ones and some small ones.
- Start with a unit lined up with one end of the object.
- Line up the units in a straight line end-to-end with no gaps.
- Do not overlap the units.
- Give the units when you say your measurement. For example, Emma said the spoon measured about 5 *paper clips* long.

Try It Together

Measure the length of a spoon using paper clips, toothpicks, or other same-size units.
Ask a partner to measure the length of the spoon. Do your measurements agree? Why?

You can use **tools** to measure length.
These tools use **standard units** that never change.
Standard units are the same for everyone.

meterstick

yardstick

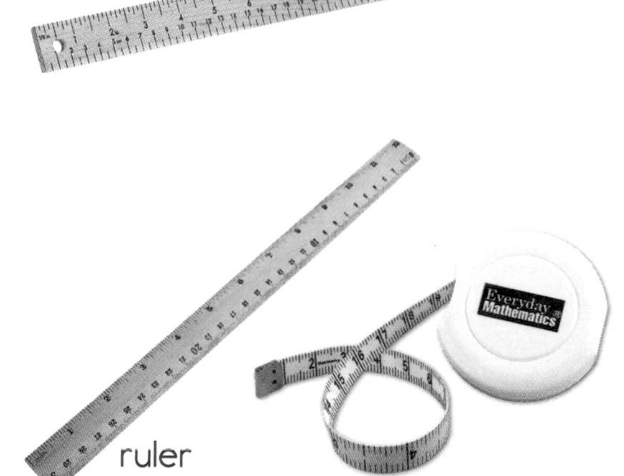

ruler

tape measure

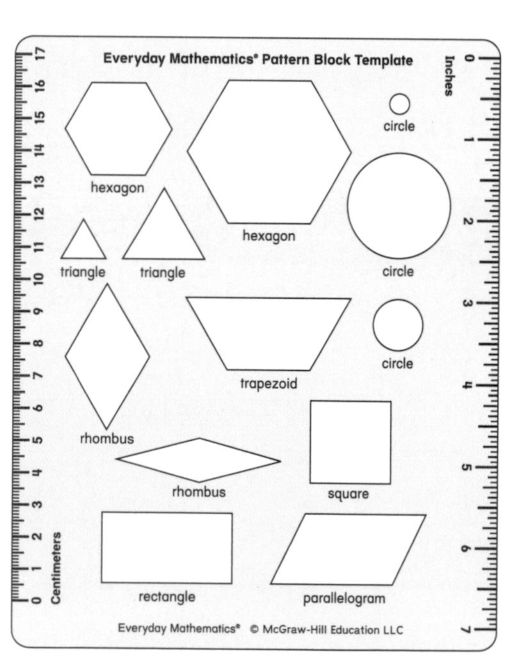

Pattern-Block Template

Note You can use a tape measure to measure the distance around objects. We call this distance the **circumference.**

An **inch** is one kind of standard unit.

Rulers are often marked with inches.

Measure to the nearest inch.

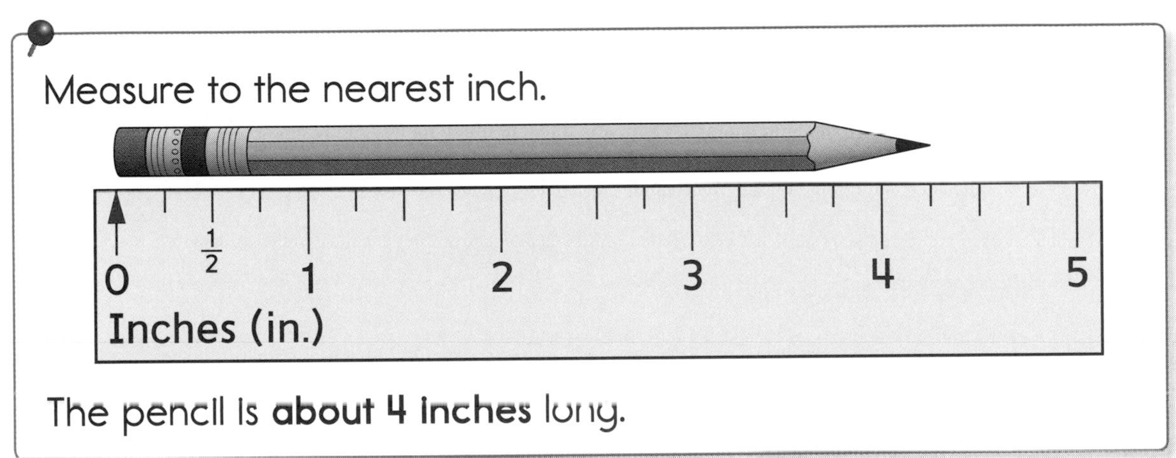

The pencil is **about 4 inches** long.

Measure to the nearest inch.

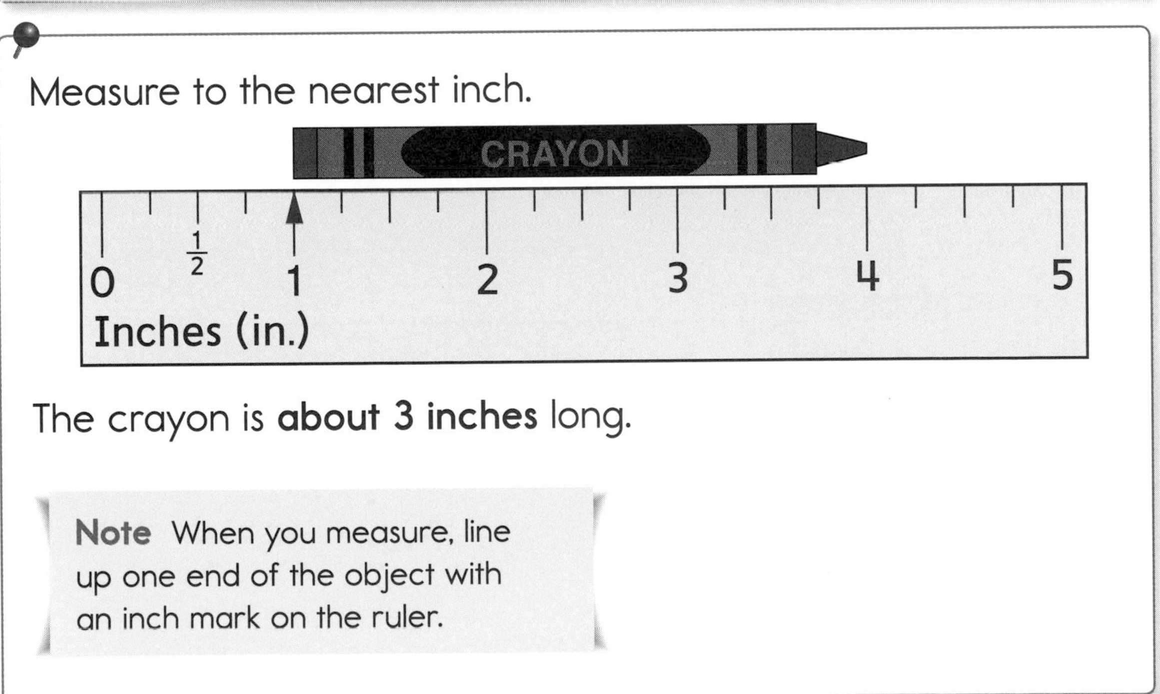

The crayon is **about 3 inches** long.

Note When you measure, line up one end of the object with an inch mark on the ruler.

You can use **personal references** to help estimate the length of objects. A personal reference can be part of your body or a familiar object.

You can use the distance from the tip of your thumb to the first joint as a personal reference for one inch. Your thumb will be slightly different than other children's, but you can use it to estimate.

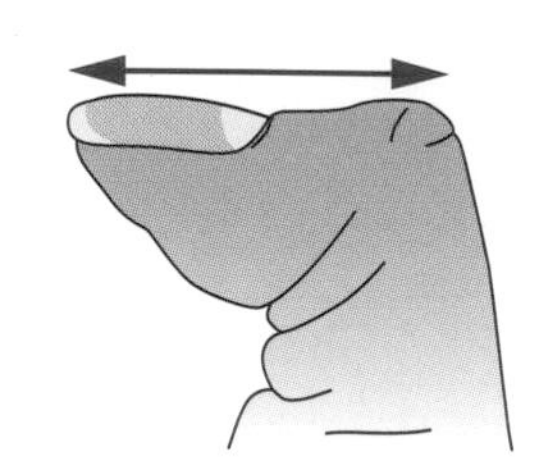

You can use your thumb to estimate the length of a key. Imagine how many times your thumb will fit along the key. A good estimate for the length of the key is about 2 inches.

Try It Together

Use your thumb to estimate the length of a pencil in inches.

A **yard** is another standard unit of measure.

A yard is 36 inches long. A good tool for measuring yards is a yardstick.

A good personal reference for estimating yards is the width of a door.

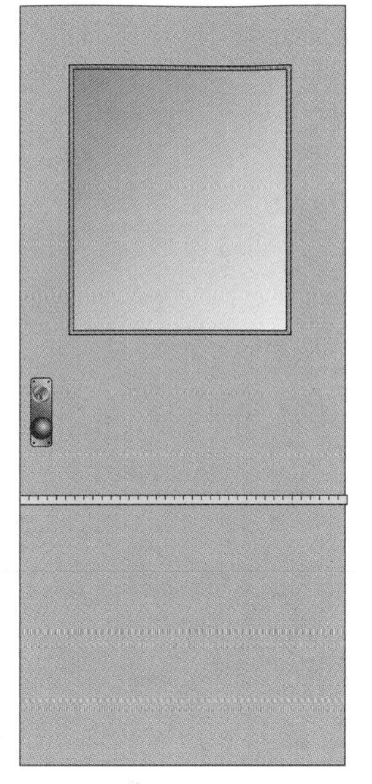

A **foot** is a standard unit of measure.

A foot is 12 inches long. Good tools for measuring feet are a yardstick or a ruler.

A good personal reference for estimating feet is the length of a piece of paper.

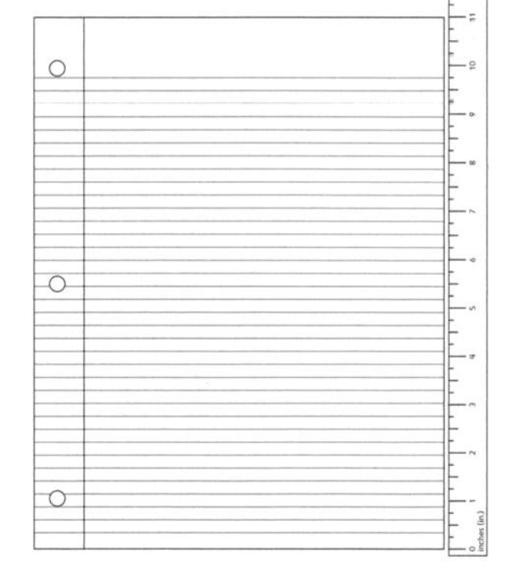

Note Inches, feet, and yards are part of the U.S. customary system of measurement.

A **centimeter** is a standard unit of measure.

Measure to the nearest centimeter.

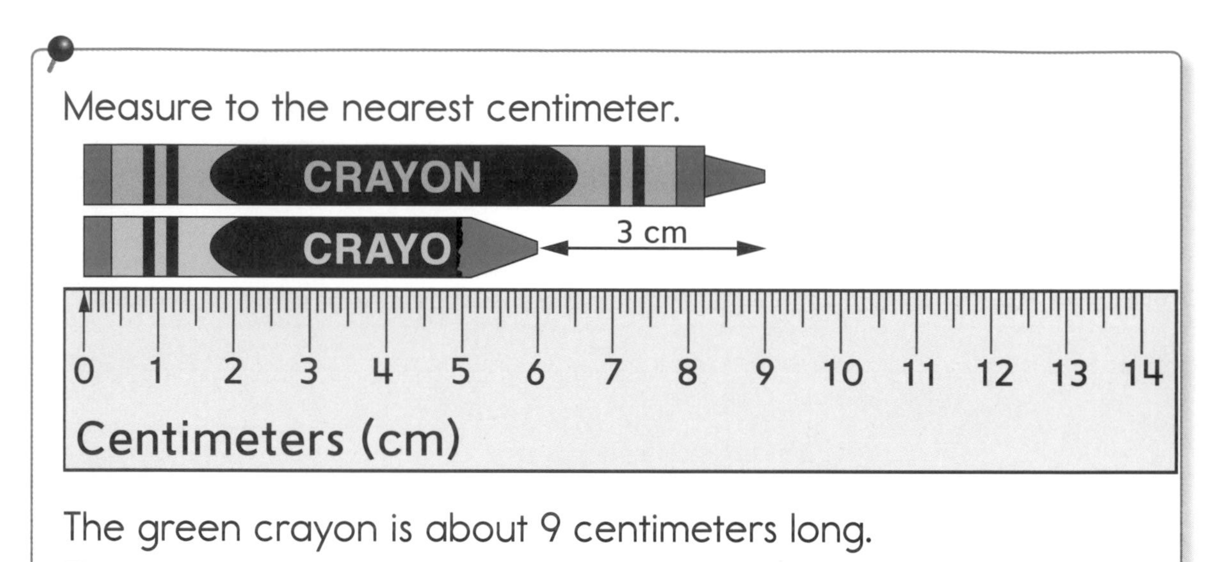

The green crayon is about 9 centimeters long.
The blue crayon is about 6 centimeters long.

The green crayon is about 3 centimeters longer than the blue crayon.

9 − 6 = 3 centimeters

A good personal reference for a centimeter is the width of your little finger.

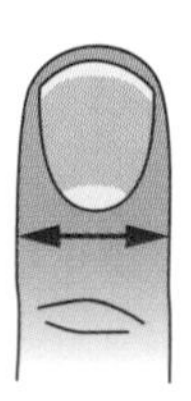

Note A **meter** is 100 centimeters. Meters and centimeters are part of the metric system.

Try It Together

Use your little finger to estimate the length of a pencil in centimeters.

Time

Read It Together

There are many ways to show **time.** One way uses hour and minute hands and numbers to show time. Another way uses only numbers.

You can use analog clocks to show time.

minute mark

hour mark and number

analog clock

You can use digital clocks to show time.

hour 8:05 minutes

Thursday, June 6

slide to unlock

digital clock

Read the hour numbers on the analog clock.

Did You Know?

The first clock with a minute hand was invented in 1577.

The digital clock was invented in 1956.

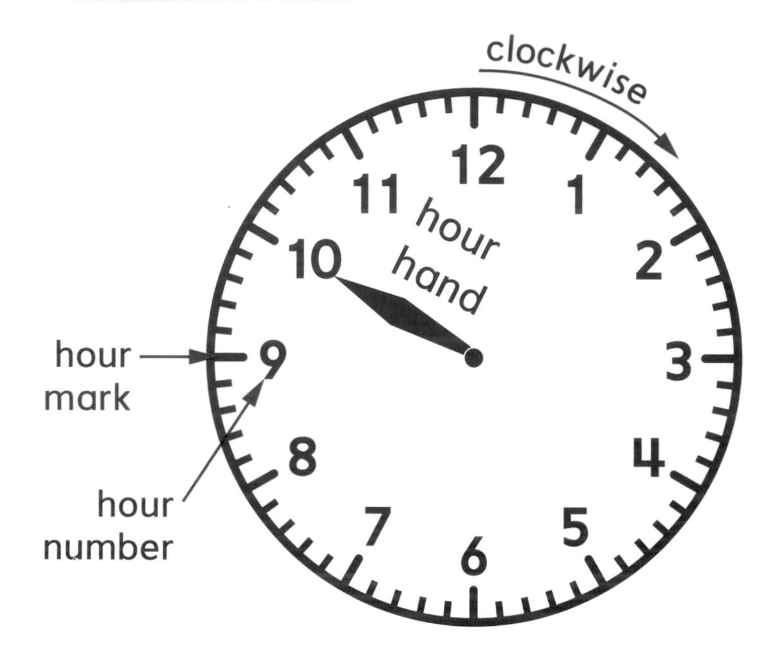

The **hour hand** shows the hour.
The hour hand is the shorter hand on a clock face.

The hour hand takes 1 hour to move from an hour mark to the next hour mark.

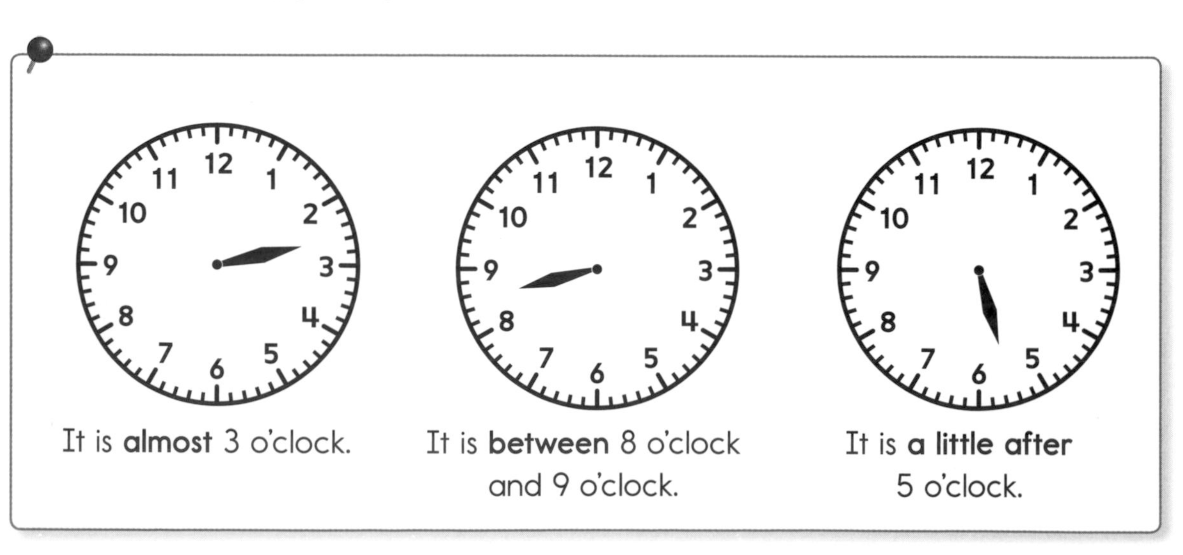

It is **almost** 3 o'clock.

It is **between** 8 o'clock and 9 o'clock.

It is **a little after** 5 o'clock.

The **minute hand** shows the minutes.

The minute hand is the longer hand on a clock face.

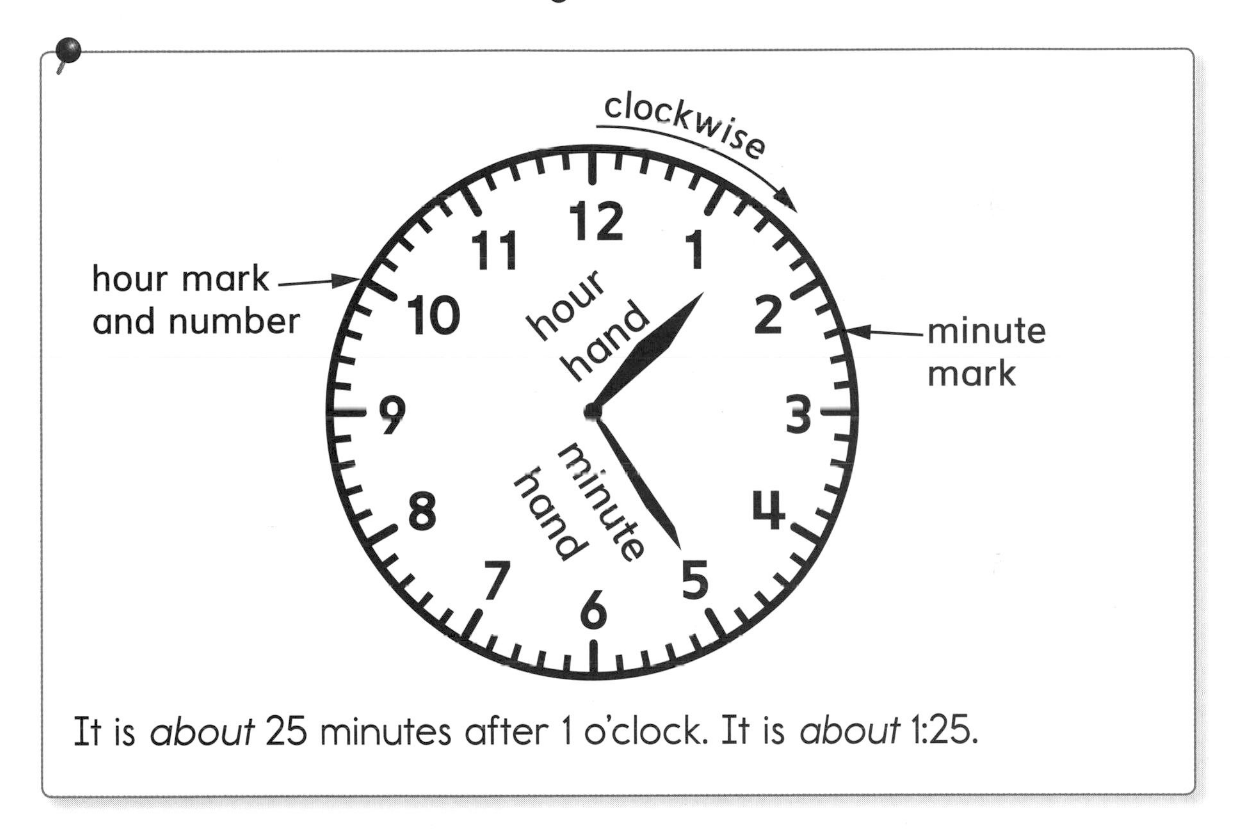

It is *about* 25 minutes after 1 o'clock. It is *about* 1:25.

The minute hand takes 1 minute to move from a minute mark to the next minute mark.

All hands on a clock move **clockwise.**

The hours from midnight to noon are A.M. hours.

12:00 A.M.
12 midnight

7:30 A.M.
half past 7

10:15 A.M.
a quarter after 10

12:00 P.M.
12 noon

The hours from noon to midnight are P.M. hours.

12:00 P.M.
12 noon

3:30 P.M.
half past 3

7:45 P.M.
a quarter to 8

12:00 A.M.
12 midnight

Money

Read It Together

We use **money** to buy things.
Here are some of the coins and bills used in the United States.

	Penny 1¢ $0.01	Nickel 5¢ $0.05
Heads or Front		
Tails or Back		
Equivalencies	1 Ⓟ	1 Ⓝ 5 Ⓟ

United States Mint image

Note Use the words *heads* and *tails* when talking about coins.

Use the words *front* and *back* when talking about bills.

How many nickels make 3 dollars?

Dime 10¢ $0.10	Quarter 25¢ $0.25	Dollar 100¢ $1.00
 LIBERTY IN GOD WE TRUST	 UNITED STATES OF AMERICA LIBERTY IN GOD WE TRUST QUARTER DOLLAR	 FEDERAL RESERVE NOTE THE UNITED STATES OF AMERICA F 00775895 H WASHINGTON, D.C. ONE DOLLAR
 UNITED STATES OF AMERICA ONE DIME	 UNITED STATES OF AMERICA QUARTER DOLLAR	 THE UNITED STATES OF AMERICA IN GOD WE TRUST ONE ONE DOLLAR
1 Ⓓ 2 Ⓝ 10 Ⓟ	1 Ⓠ 5 Ⓝ 25 Ⓟ	1 $1 4 Ⓠ 10 Ⓓ 20 Ⓝ 100 Ⓟ

You can use **dollars-and-cents notation** to write money amounts.

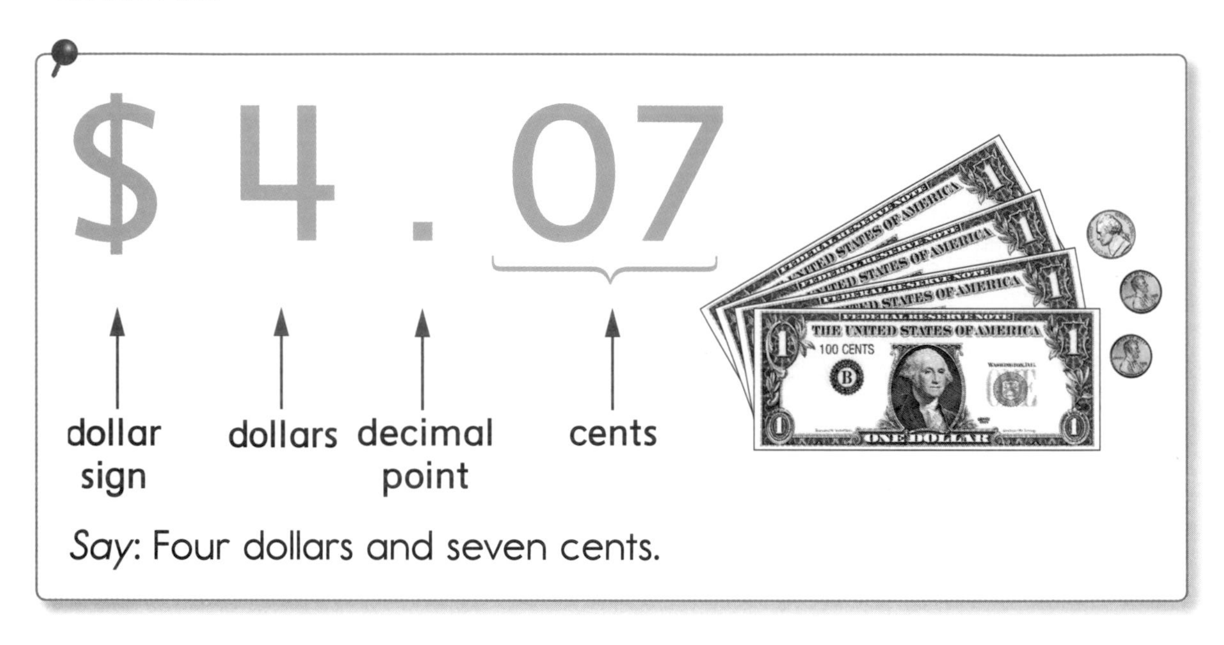

You can use symbols to show money amounts.

\$1 \$1 \$1 \$1 Ⓝ Ⓟ Ⓟ

Try It Together

Play the *Exchange Game* using pennies, dimes, and dollars. The rules are on pages 146–148.

Tally Charts and Line Plots

Read It Together

Information someone has collected is called **data.**

A **tally chart** is one way to organize data.

Teeth Lost in Mr. Alan's Class

Number of Teeth Lost	Number of Children
0	𝍸
1	𝍸 //
2	///
3	////
4	/
5	

This tally chart shows the number of teeth lost by the children in Mr. Alan's class.

Remember: Every fifth tally mark goes across a group of four tally marks.

𝍸

There are 3 tally marks to the right of 2.
This means that 3 children lost 2 teeth.

A **line plot** is another way to organize data.

Teeth Lost in Mr. Alan's Class

Number of Children

	X				
	X				
X	X				
X	X		X		
X	X	X	X		
X	X	X	X		
X	X	X	X	X	
0	1	2	3	4	5

Number of Teeth Lost

This line plot shows the same data as the tally chart on page 113.

There are 3 Xs above 2.
This means that 3 children lost 2 teeth.

How many teeth have you lost?

Try It Together

How many children did not lose any teeth?

Graphs

Read It Together

A **picture graph** uses a picture or a symbol to show data.

Books Read in Ms. Cook's Class

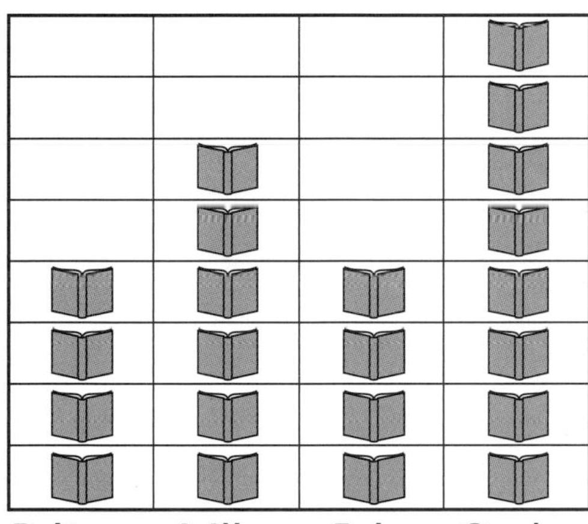

KEY: Each = 1 Book

There are 4 above Britney.
This means that Britney read 4 books.

There are 6 above Mike.
This means that Mike read 6 books.
Mike read 2 more books than Britney.

Try It Together

How many more books did Carlos read than John?

A **bar graph** uses bars to show data. This bar graph shows the same data as the picture graph on page 115.

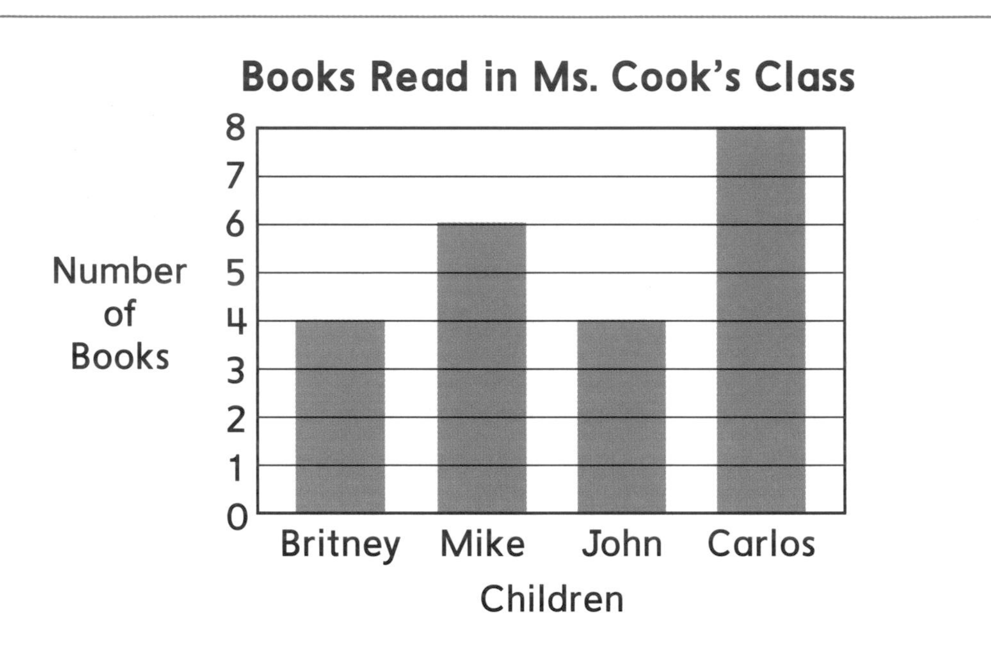

The bar above Carlos shows that he read 8 books. It is the tallest bar, and it shows that he read the most books.

The bar above Mike shows that he read 6 books. This shows that he read 2 fewer books than Carlos.

Try It Together

1. How many books did Mike and Carlos read altogether?
2. How many more books did Mike read than John?

Measures All Around: Animals and Tools

Have you ever wondered how large animals are measured? What about tiny or dangerous animals? Scientists use many different tools to measure animals.

A fish biologist uses a ruler to measure the length of this minnow. He can estimate the fish's age if he knows the length.

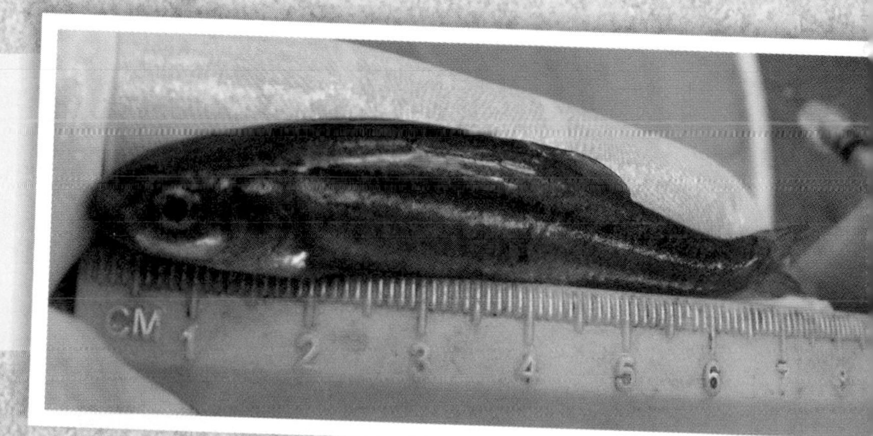

(t)Richard Walker, (b)Doug Helton, NOAA/NOS

This oceanographer uses a caliper to measure the length of the coral to find out how much it has grown. A caliper measures things from end to end.

This coral reef is located in the South Pacific Ocean near the islands of American Samoa.

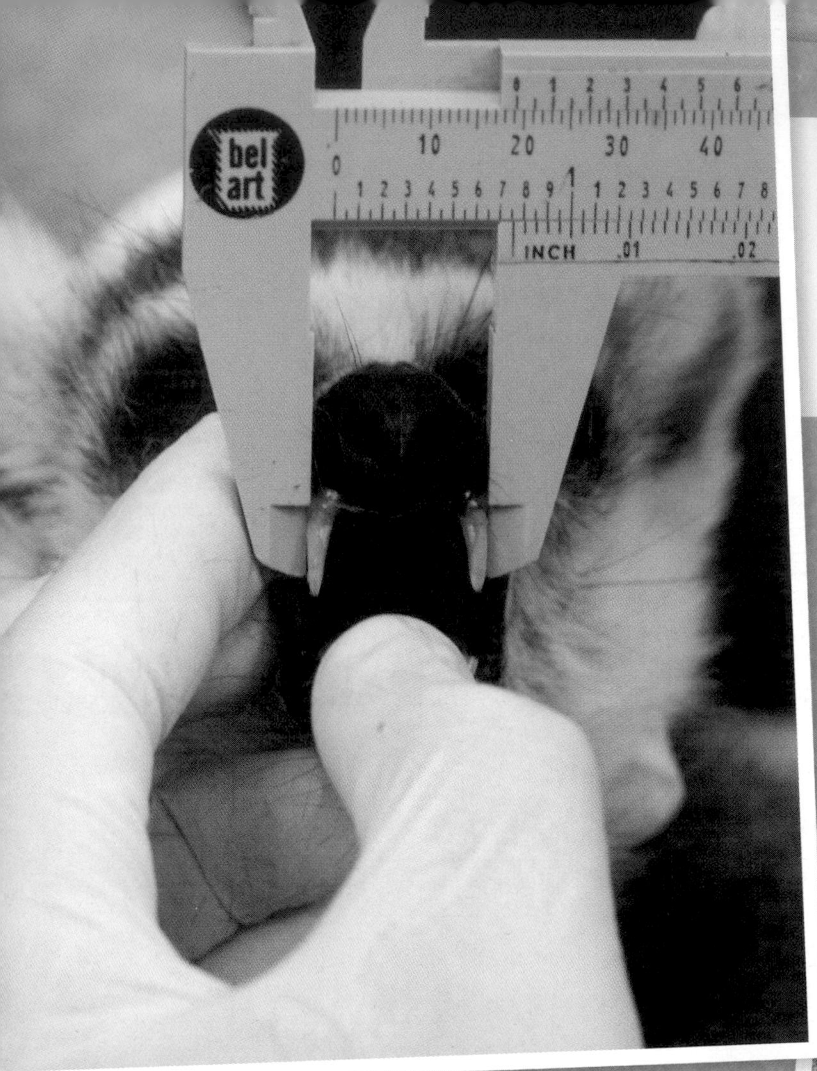

A zoologist is a scientist who studies animals. A zoologist uses a caliper to measure the distance between a lemur's canine teeth.

Lemurs are found in the wild only on the island of Madagascar.

A zoologist is using a special thermometer to measure the temperature of this sloth.

Sloths live in trees in the rain forests of Central and South America.

This zoologist measures the kangaroo's resting heart rate. He listens through a stethoscope and counts how many times he hears the heart beat in one minute.

This is a stethoscope.

Do you think the kangaroo's heart rate gets faster or slower when it hops around?

Kangaroos are found in the wild only in Australia.

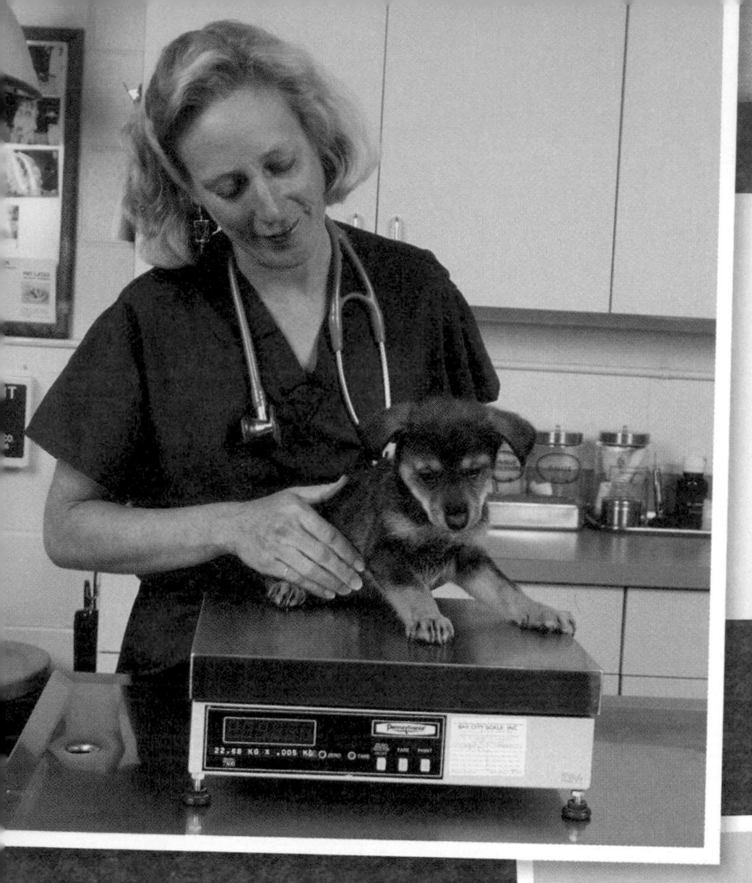

This veterinarian uses a digital scale to weigh a puppy. This scale measures weight in pounds and ounces. Why might veterinarians want to know the weight of a young animal?

These scientists are weighing a chicken using a spring scale. What animals are heavier than a chicken?

What tools do *you* use to measure?

Geometry

2-Dimensional Shapes

Read It Together

Flat shapes are 2-dimensional shapes. You can draw 2-dimensional shapes on paper, but you can't hold them in your hand.

This 2-dimensional shape is a **polygon.**
Polygons are made of straight sides called **line segments.**

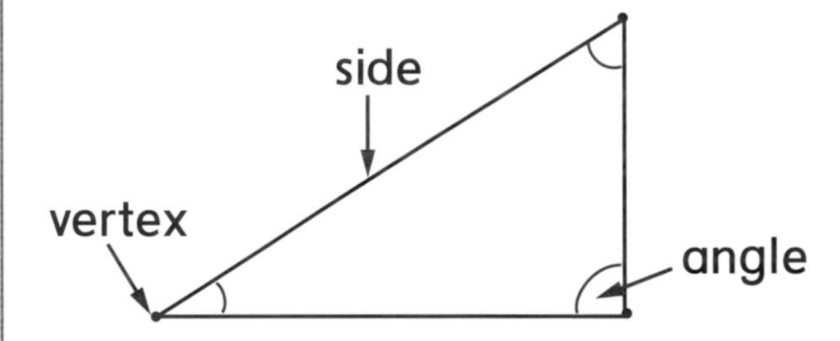

Note A point where two sides meet is called a **vertex.** The plural of *vertex* is *vertices.*

This polygon has 3 **sides,** 3 **angles,** and 3 **vertices.**

This 2-dimensional shape is also a polygon.

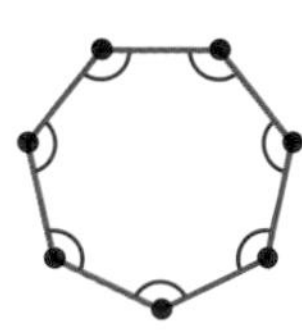

Count the sides, angles, and vertices of this polygon.

Here are more 2-dimensional shapes.

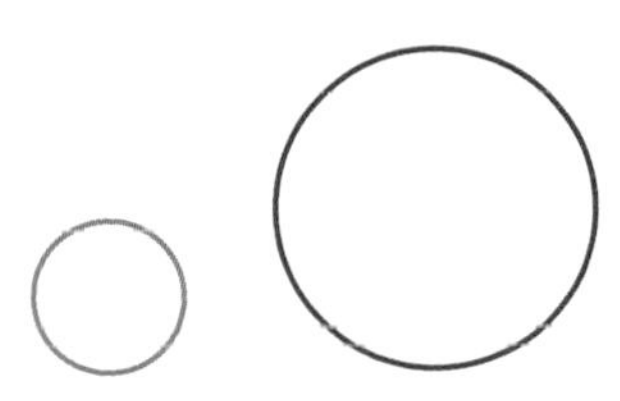

A **circle** has no straight sides.

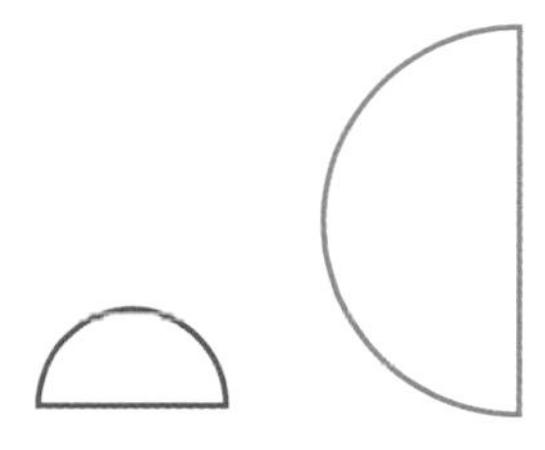

A **half circle** has 1 straight side.

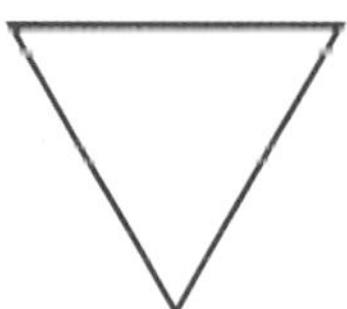

A **triangle** has 3 sides.

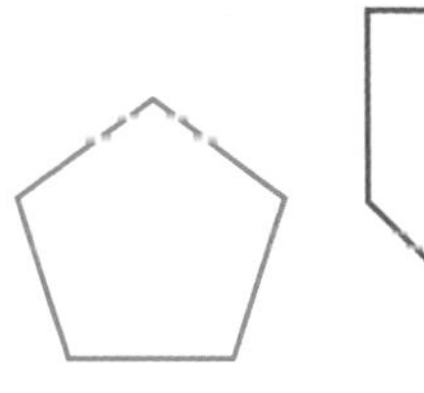

A **pentagon** has 5 sides.

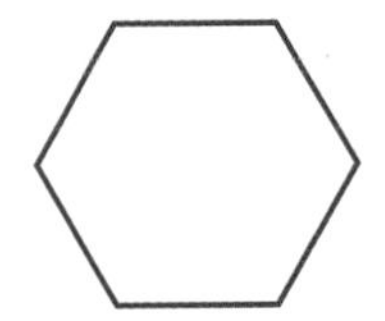

A **hexagon** has 6 sides.

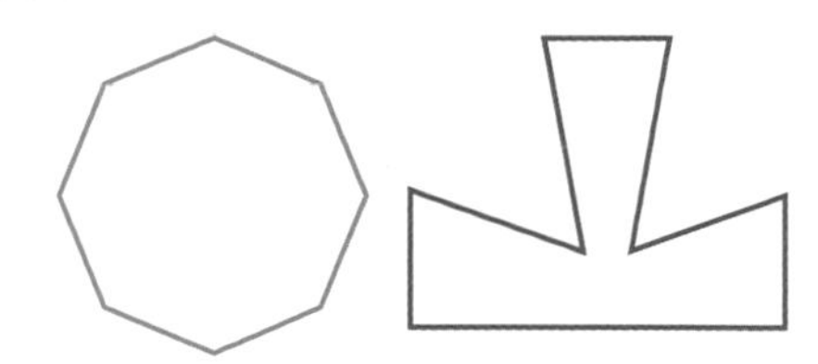

An **octagon** has 8 sides.

Try It Together

Choose one of the 2-dimensional shapes on this page.
Describe it to a partner without saying the shape's name.
Can your partner tell what shape you are describing?

These 2-dimensional shapes are **polygons.**

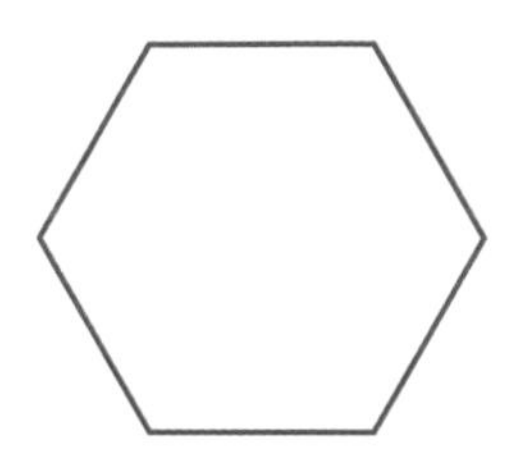
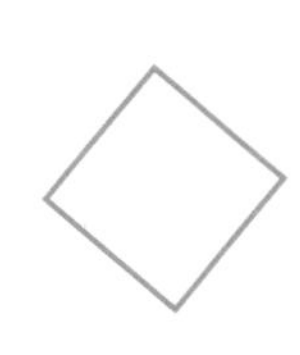
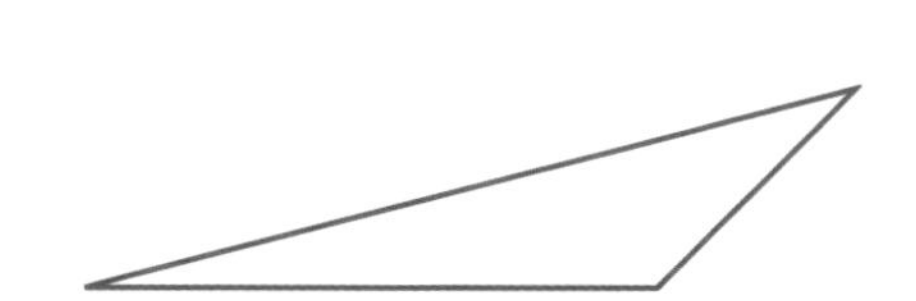
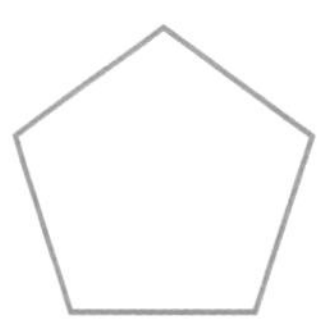
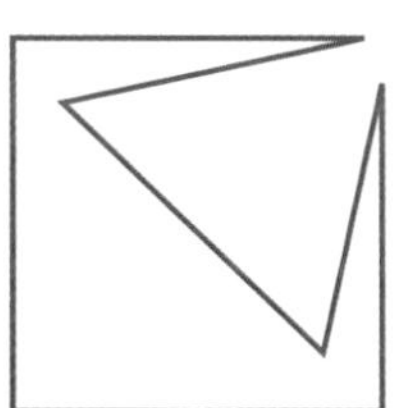

These 2-dimensional shapes are *not* polygons.

The sides cross.

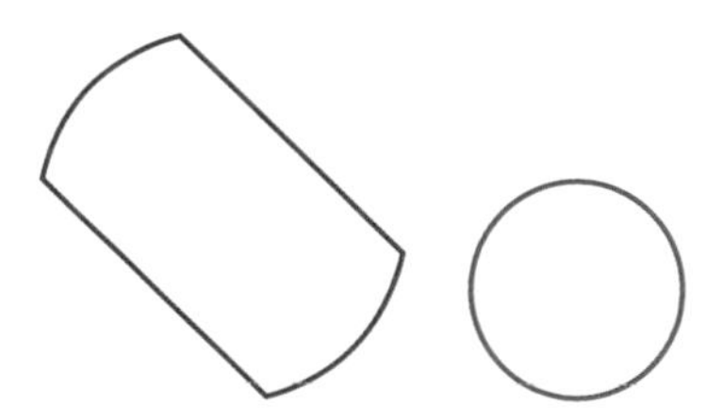

Some sides are curved.

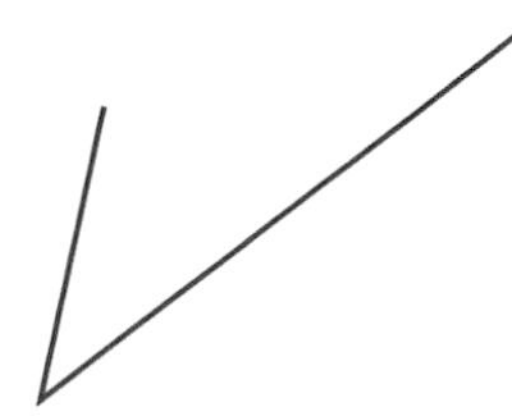

There are not 3 or more sides.

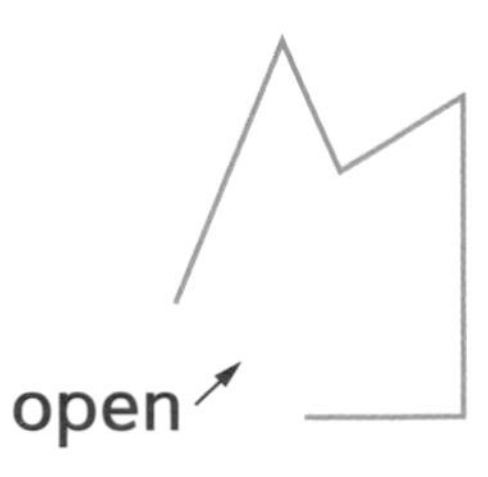

The sides are not closed.

Shapes have different **attributes** such as number of sides, number of vertices, size, and color. Every polygon has certain attributes that make it a polygon.

Defining Attributes of Polygons	Nondefining Attributes of Polygons
All polygons . . . • are flat • have 3 or more sides • have straight sides that do not cross • have sides that meet at a vertex • are closed	**Some polygons might . . .** • be different sizes • be different colors • be turned in different directions

Choose a shape from page 124 that is a polygon. Show how you know that it has the defining attributes of a polygon.

Some 2-dimensional shapes have **parallel** sides.

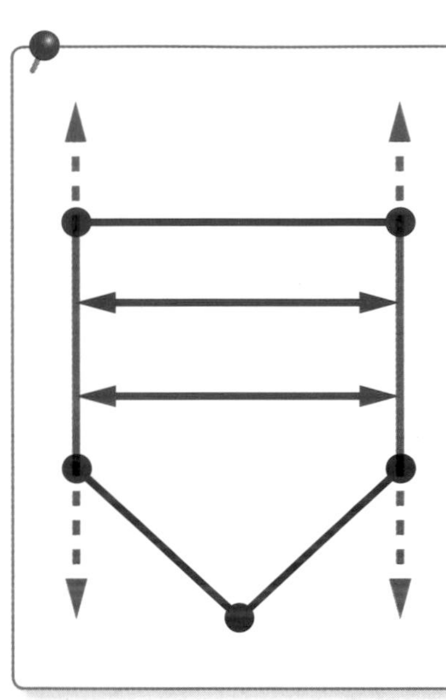

This polygon has 5 straight sides, 5 angles, and 5 vertices. The sides meet at a vertex.

Two sides of the shape are **parallel.** Sides of a shape are parallel if they are always the same distance apart. If you extend the line segments in both directions, the lines will always be the same distance apart.

This shape has two pairs of parallel sides.

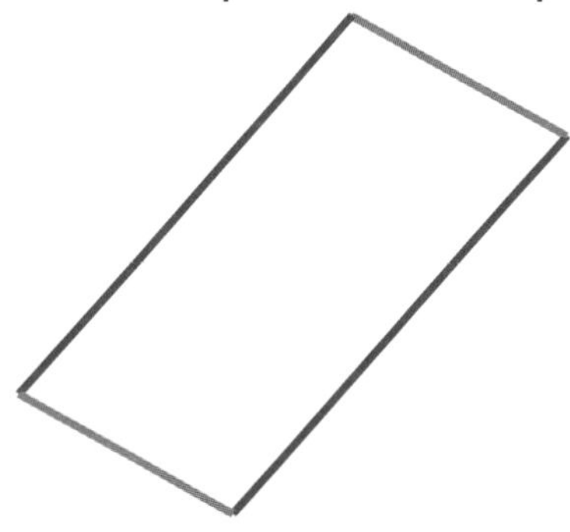

Note You can check to see if sides of a shape are parallel. Take a straight object, such as a pencil, and place it on top of one side. Move the pencil toward the other side to see if it can slide on top of the other side without turning.

These polygons are **quadrilaterals.**
Quadrilaterals are also called **quadrangles.**

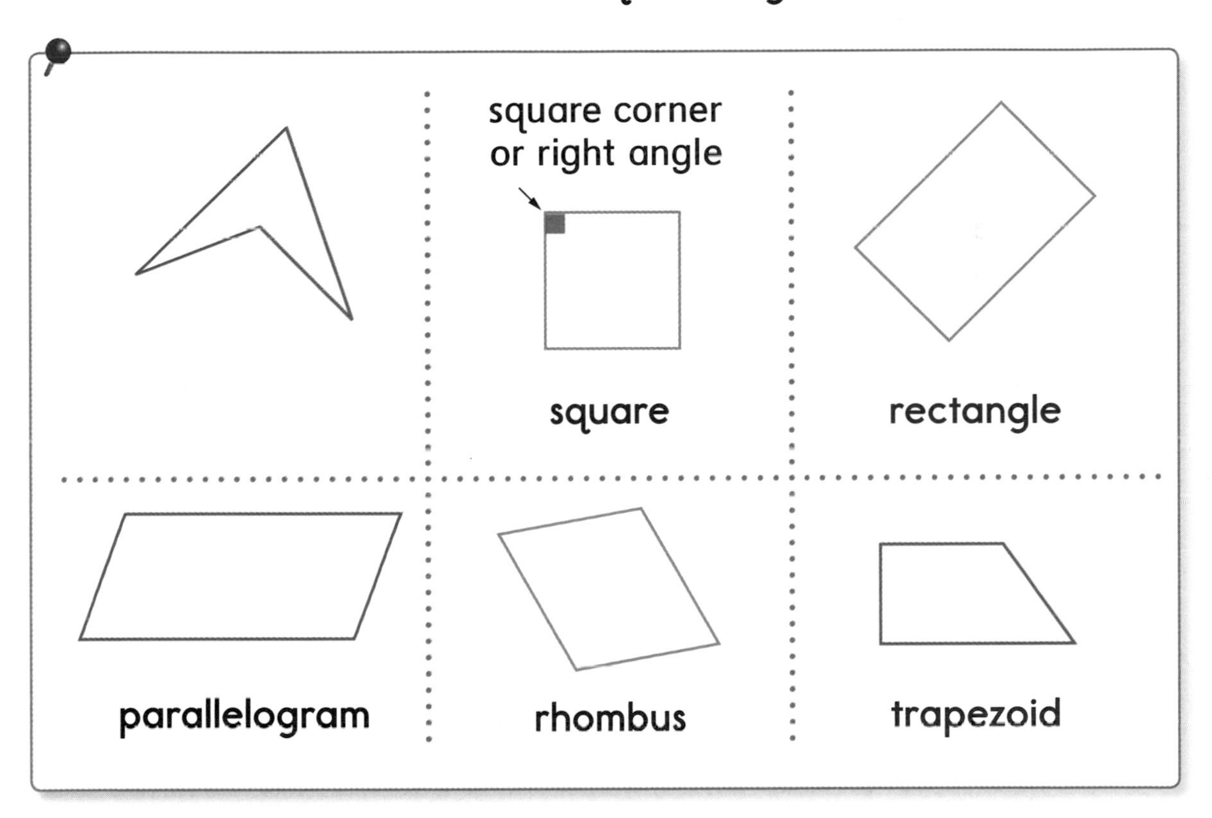

Defining Attributes of Quadrilaterals	Nondefining Attributes of Quadrilaterals
All quadrilaterals . . .	**Some quadrilaterals might . . .**
• have 4 straight sides • have 4 vertices and 4 angles	• be turned in different directions • have parallel sides

These quadrilaterals are **rectangles.**

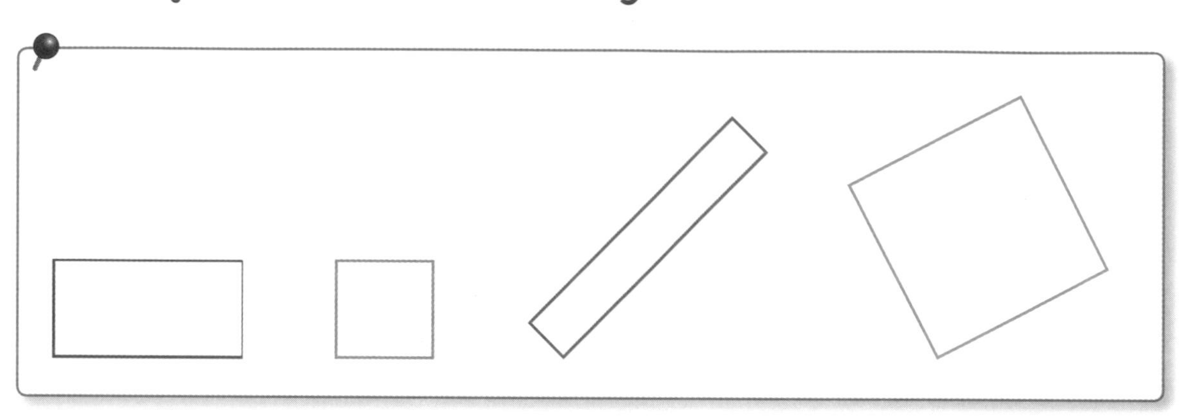

Defining Attributes of Rectangles	Nondefining Attributes of Rectangles
All rectangles . . . • have 4 straight sides • have 4 square corners or right angles • have 2 pairs of equal-length sides • have 2 pairs of parallel sides	**Some rectangles might . . .** • be different sizes • be turned in different directions • have all sides the same length

You can put shapes together to make new shapes.

Two triangles can make a rectangle.

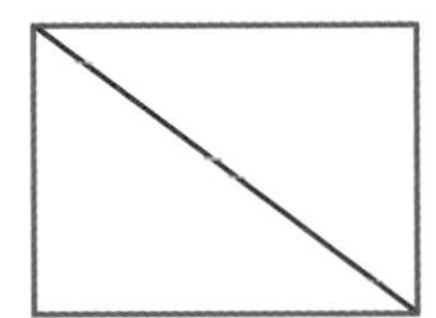

You can add another triangle to make a trapezoid.

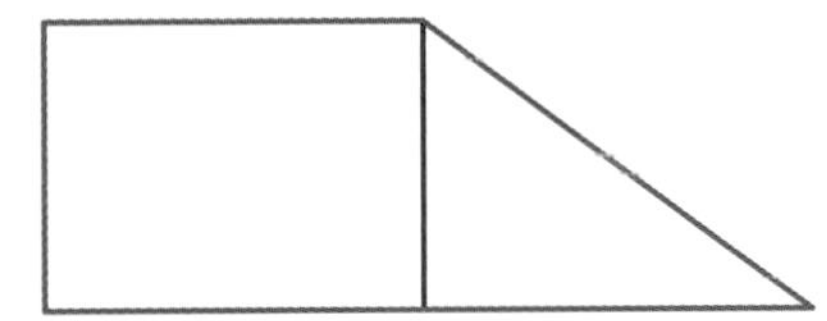

You can put shapes together to make other interesting shapes.

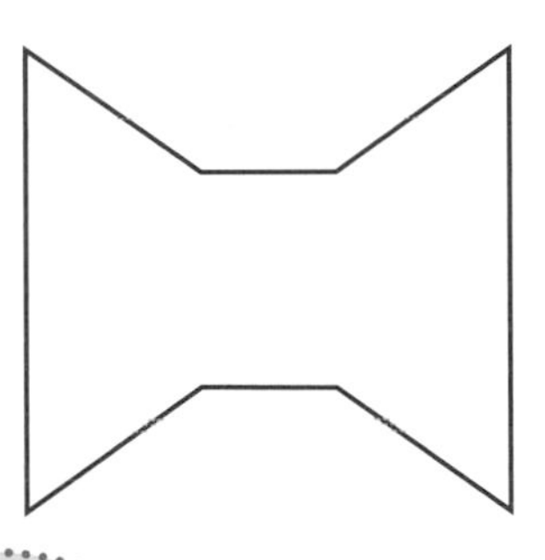

What shapes can you put together to make the red shape?

What shapes can you put together to make the green shape?

Try It Together

Look back at the shapes on pages 123, 124, 127, and 128. Draw a new shape using two or more of the shapes.

Sometimes you want to know the amount of surface inside a shape. The amount of surface inside a shape is the **area** of the shape.

Area is measured in square units.

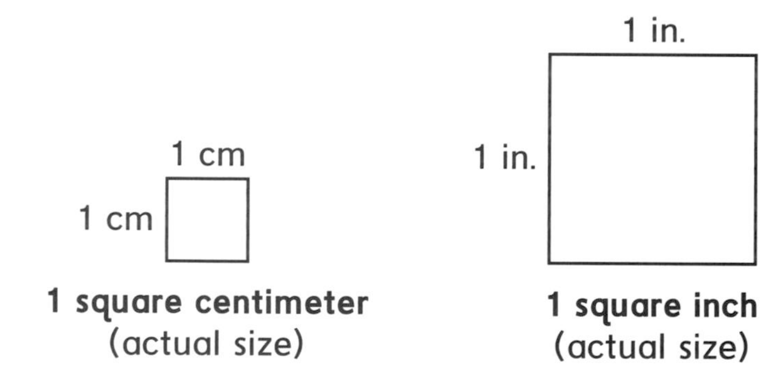

1 square centimeter
(actual size)

1 square inch
(actual size)

You can count the squares to find the area of this hexagon.

Each square is 1 square centimeter.

There are 6 squares in the hexagon.

The area of the hexagon is 6 square centimeters.

You can cover a shape with unit squares to find the area. The area is the number of squares.

Count the squares to find the area.

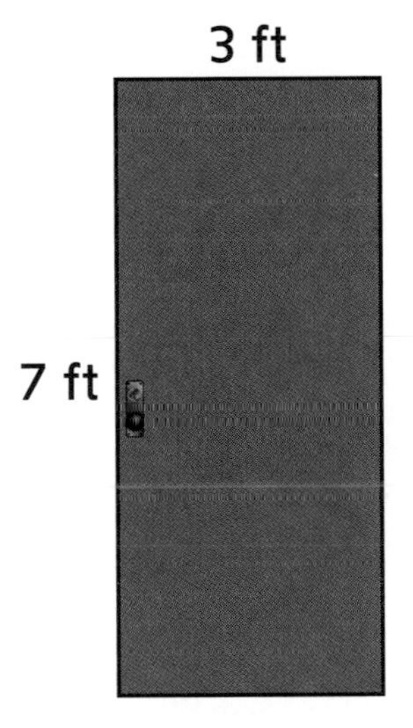

1	2	3
4	5	6
7	8	9
10	11	

1	2	3
4	5	6
7	8	9
10	11	12
13	14	15
16	17	18
19	20	21

The area of the front of this door is 21 square feet.

Try It Together

Draw a rectangle that has an area of 12 square units. Compare your shape with a partner's shape.

Fractions and Equal Parts

Read It Together

A whole object can be divided into **equal parts.**

a whole tortilla

the whole tortilla divided into 2 equal shares

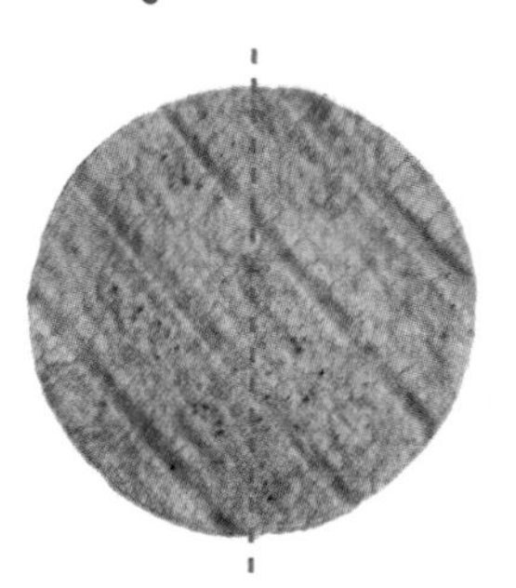

Here are ways you can describe 1 share of the tortilla:

1 half one-half 1 out of 2 equal shares

Here are some ways you can describe both shares:

the whole tortilla two-halves 2 out of 2 shares

Note You can show that 2 parts of a circle are equal by folding the circle so that the parts lay on top of one another. If the parts match, they are the same size.

a whole paper square

the whole square divided into 4 equal parts in two ways

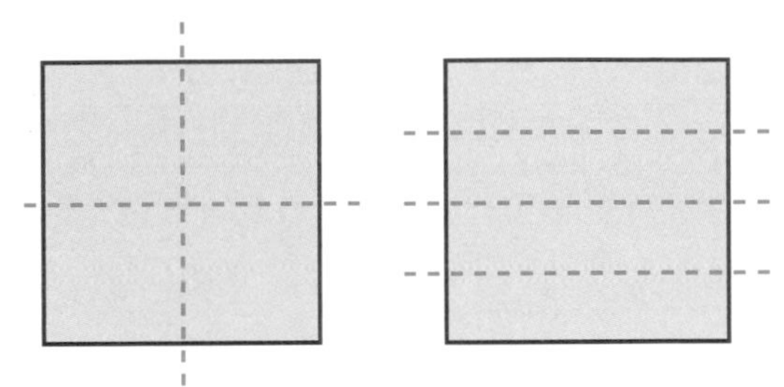

Here are ways to name one part of the square:

1 fourth　　a quarter　　1 out of 4 equal parts

Here are ways you can describe all the parts:

the whole square　　four-fourths　　4 out of 4 parts

a whole cracker

the whole cracker divided into 3 equal shares in two ways

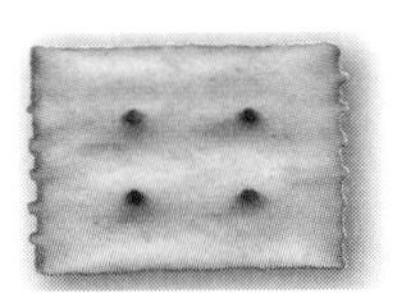

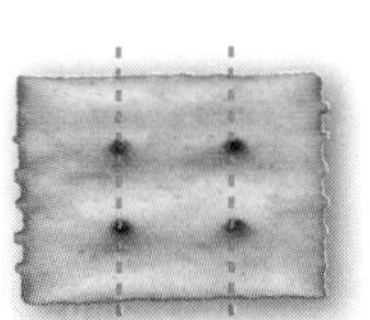

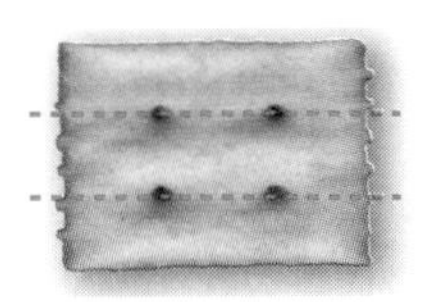

Here are ways to name one share of the cracker:

1 third　　one-third　　1 out of 3 equal shares

Here are ways you can describe all the shares:

the whole cracker　　3 thirds　　3 out of 3 shares

3-Dimensional Shapes

Read It Together

A 3-dimensional shape takes up space.
Objects you can hold are 3-dimensional.

Some 3-dimensional shapes are **rectangular prisms.**

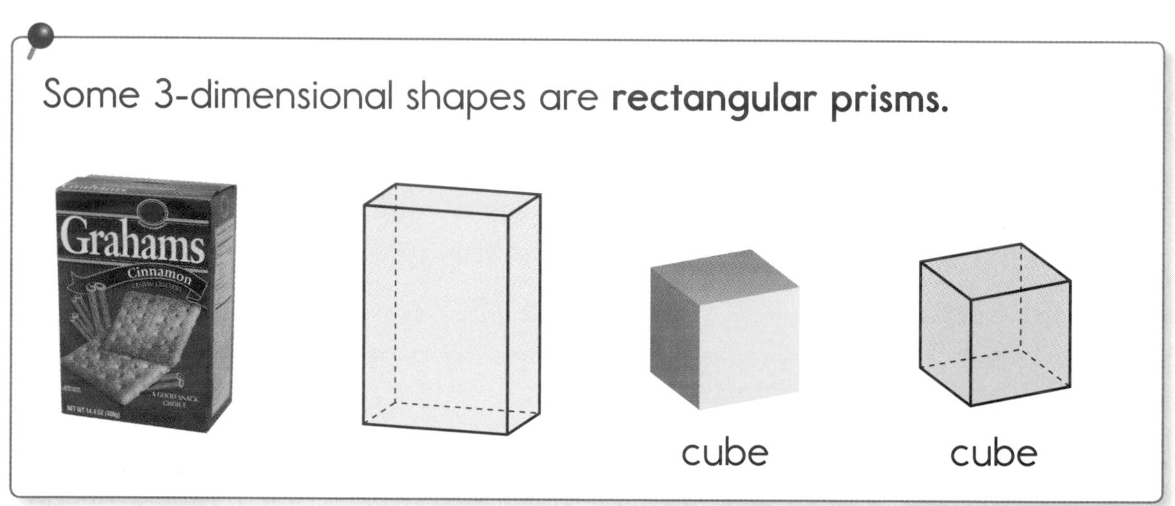

Some 3-dimensional shapes are **cylinders.**

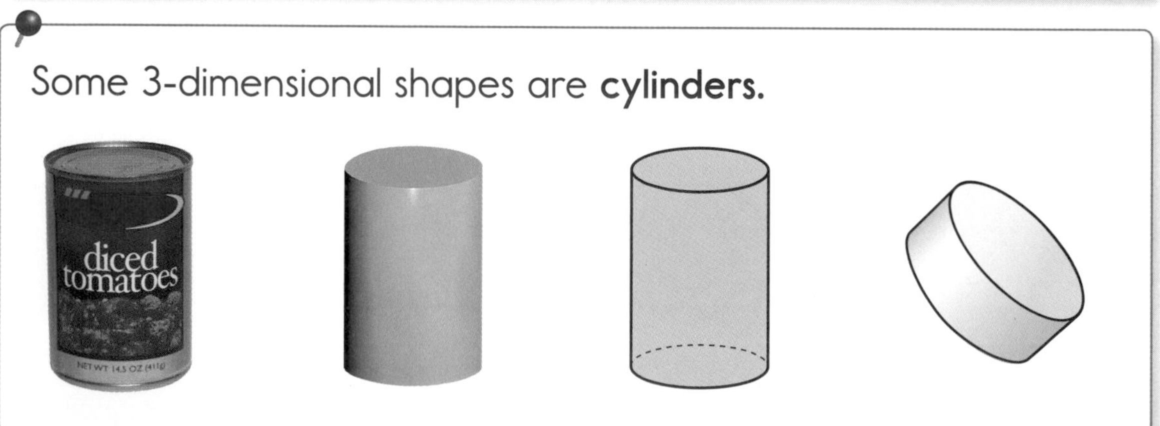

The dotted lines in the drawings help you imagine the parts of the shapes you can't see in the pictures.

Some 3-dimensional shapes are **cones.**

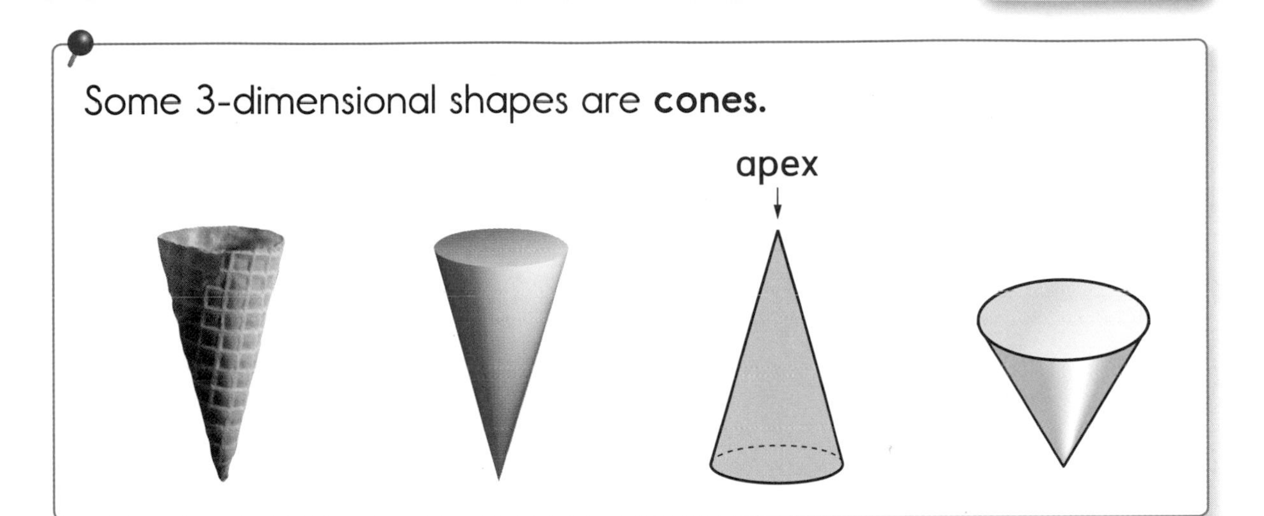

Some 3-dimensional shapes are **spheres.**

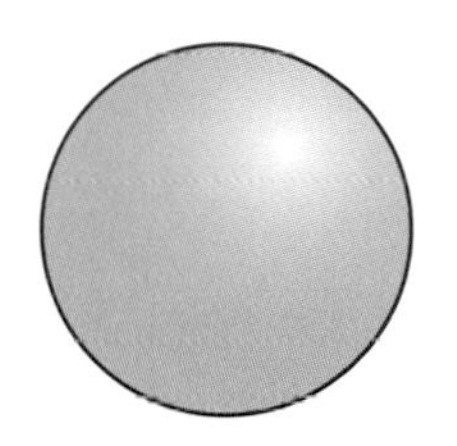

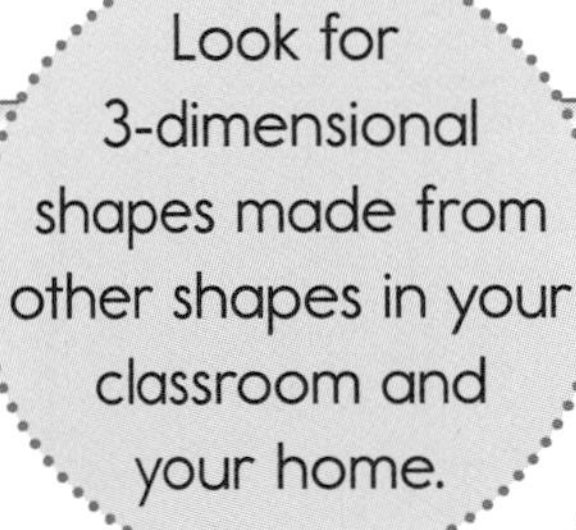

You can make 3-dimensional shapes from other 3-dimensional shapes.

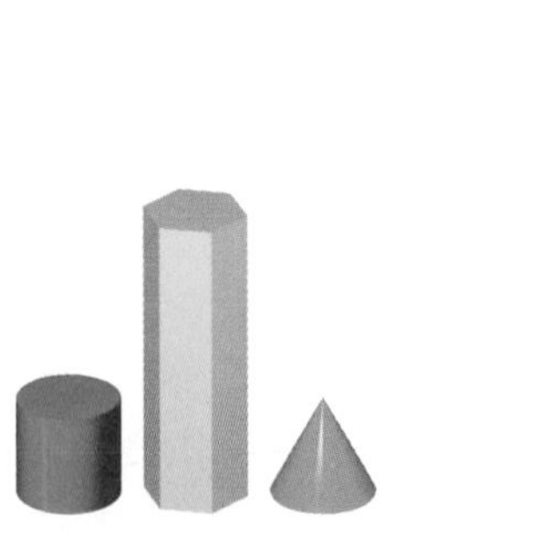
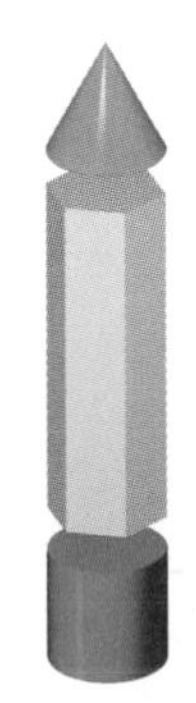
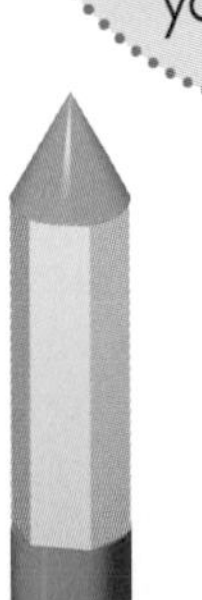

Some 3-dimensional shapes have faces, edges, and vertices.

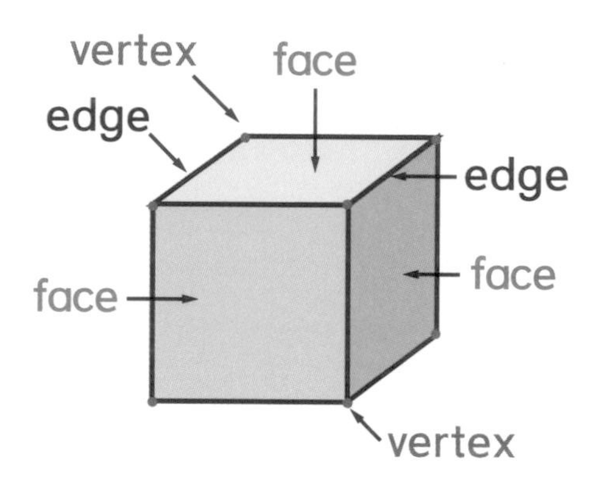

A cube has 6 faces, 12 **edges,** and 8 vertices. All the faces are squares.

Here is one way to draw a cube.

Draw a square for the front face.

Draw 3 short lines for edges.

Connect the 3 edges.

Games

fStop/PunchStock

Addition/Subtraction Spin

Materials
- ☐ 1 *Addition/Subtraction Spin* Spinner
- ☐ 1 paper clip
- ☐ 1 pencil
- ☐ 1 die marked with + 10, − 10, + 100, and − 100
- ☐ 1 calculator
- ☐ 2 sheets of paper

Players 2

Skill Mentally adding and subtracting 10 and 100

Object of the Game To have the larger total.

Directions

1. Players take turns being the "Spinner" and the "Checker."
2. The Spinner uses a pencil and a paper clip to make a spinner.

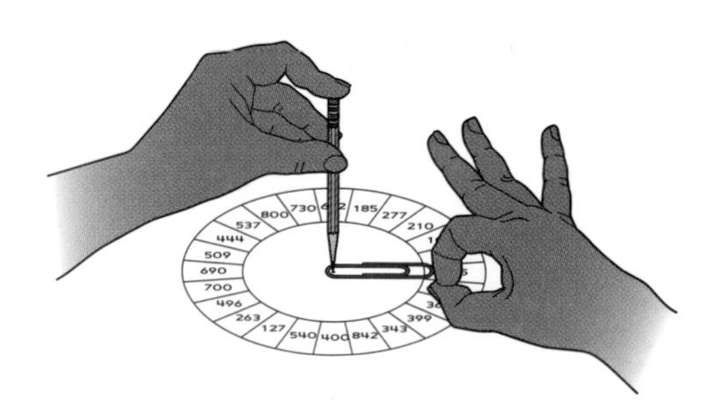

3. The Spinner spins the paper clip and writes the number that the paper clip points to. If the paper clip points to more than one number, the Spinner writes the smaller number.

4. The Spinner rolls the die and records what is shown on the top.
5. The Spinner adds or subtracts 10 or 100 to solve the problem and writes the answer. The Checker checks the answer by using a calculator.
6. If the answer is correct, the Spinner circles it. If the answer is incorrect, the Spinner corrects it but does not circle it.
7. Players switch roles. They stop after they have each played 5 turns. Each player uses a calculator to find the total of his or her circled scores.
8. The player with the larger total wins.

Vern spins 554 and rolls $-$ 10. He writes $554 - 10 = 544$.

Jane checks it on a calculator and agrees it is correct. Vern circles 544.

Before and After

Materials ☐ number cards 0–10 (4 of each)
Players 2
Skill Identifying numbers that are 1 less or 1 more than a given number
Object of the Game To have fewer cards.

Directions

1. Shuffle the cards.
2. Deal 6 cards to each player.
3. Place 2 cards number-side up on the table. Put the rest of the deck number-side down.
4. Players take turns. When it is your turn:
   - Look for any numbers in your hand that come *1 before* or *1 after* one of the numbers on the table. Put your numbers on top of the numbers on the table. Play as many cards as you can.
   - Take cards from the deck so that you have 6 cards again.

- If you can't play any cards when your turn begins:
  Place 2 cards from the deck
  number-side up on top of the 2 cards
  on the table.
  Try to play cards from your hand again.
  If you still can't play your cards, your turn is over.

5. The game is over when:
   - All the cards have been taken from the deck.
   - No one can play any more cards.
6. The player holding fewer cards wins.

Sally gives 6 cards to Ricky and 6 cards to herself. Then she turns over a 9 card and a 2 card and places them on the table.

Ricky puts an 8 card on top of the 9 card and says, "8 is before 9." Then Ricky takes another card from the deck.

The Difference Game

Materials
- ☐ number cards 1–10 (4 of each)
- ☐ 40 pennies
- ☐ 1 sheet of paper labeled "Bank"

Players 2

Skill Finding differences

Object of the Game To take more pennies.

Directions

1. Shuffle the cards. Place the deck number-side down on the table.
2. Put 40 pennies in the bank.
3. To play a round, each player:
   - Takes 1 card from the top of the deck.
   - Takes the same number of pennies from the bank as the number shown on the card.
4. Find out how many more pennies one player has than the other. Pair as many pennies as you can.

5. The player with more pennies keeps the extra pennies.
   The rest of the pennies go back into the bank.
6. The game is over when there are not enough pennies in the bank to play another round.
7. The player with more pennies wins the game.

Amy draws an 8. She takes 8 pennies from the bank.

John draws a 5. He takes 5 pennies from the bank.

Amy and John pair as many pennies as they can.

Amy has 3 more pennies than John. She keeps the 3 extra pennies and returns 5 of her pennies to the bank.
John returns his 5 pennies to the bank.

Amy's Card 8

John's Card 5

Amy keeps the difference – 3 pennies.

United States Mint image

The Digit Game

Materials ☐ number cards 0–9 (4 of each)
Players 2
Skill Comparing 2-digit numbers
Object of the Game To collect more cards.

Directions

1. Shuffle the cards.
   Place them number-side down on the table.
2. Each player turns over 2 cards and makes the largest 2-digit number with his or her cards.
3. Each player says his or her number and what each digit represents.
4. The player with the larger number takes all the cards.
5. The game ends when no cards are left.
6. The player with more cards wins.

Tina draws a 5 and a 3. She makes the number 53.
Raul draws a 1 and a 4. He makes the number 41.

Tina

Tina's cards are a 5 and a 3.

Tina makes the number 53.

Tina says, "Fifty-three has 5 tens and 3 ones."

Raul

Raul's cards are a 1 and a 4.

Raul makes the number 41.

Raul says, "Forty-one has 4 tens and 1 one."

Tina's number is larger than Raul's number.
Tina takes all 4 cards.

Other Ways to Play

Make 3-Digit Numbers: Each player draws 3 cards and uses them to make the largest 3-digit number possible. Play continues as in the regular game.

Fewer Cards Win: The player with fewer cards wins.

The Exchange Game

Materials
- ☐ Place-Value Mat
- ☐ base-10 blocks (20 longs and 30 cubes)
- ☐ die

Players 2

Skill Place value

Object of the Game To get 10 longs.

Directions

1. Place the base-10 blocks between you and your partner.
2. Take turns rolling the die. Take the blocks as shown in the table.
3. Put the blocks on your Place-Value Mat.
4. When you can, exchange 10 cubes for 1 long.
5. The first player to get 10 longs wins.

If you roll:	then take:

	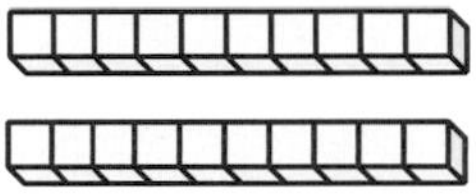

Other Ways to Play

Use a Flat: Take the number of cubes shown on the die. When you can, exchange 10 cubes for 1 long and 10 longs for 1 flat. The first player to get a flat wins.

Use Bills: Use $1, $10, and $ 100 bills. Take the number of $1 bills shown on the die. When you can, exchange ten $1 bills for one $10 bill and ten $10 bills for one $100 bill. The first player to get a $100 bill wins.

Use Pennies and Dimes: Put 30 pennies and 5 dimes on a sheet of paper labeled "Bank." Take the number of pennies shown on the die. When you can, exchange 10 pennies for 1 dime. Play until no dimes are left in the bank. The player with more dimes wins.

Use Pennies, Dimes, and Dollars: Put 1 dollar, 20 pennies, and 20 dimes on a sheet of paper labeled "Bank." Take the number of pennies shown on the die. When you can, exchange 10 pennies for 1 dime and 10 dimes for 1 dollar. The first player with 1 dollar wins.

Use Pennies, Nickels, Dimes, and Quarters: Put 20 pennies, 10 nickels, 10 dimes, and 10 quarters on a sheet of paper labeled "Bank." Take the number of pennies shown on the die. When you can, exchange 5 pennies for 1 nickel, 2 nickels for 1 dime, and 2 dimes and a nickel for 1 quarter. Play until no more quarters are left in the bank. The player with the most money wins.

High Roller

Materials ☐ 2 dice
☐ *High Roller* Record Sheet
Players 2 or more
Skill Adding numbers
Object of the Game To have the larger sum.

Directions

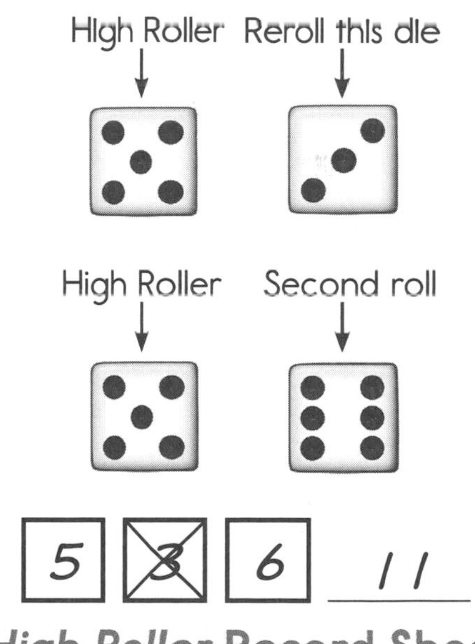

High Roller **Record Sheet**

1. Take turns. Roll both dice.
2. Record your first roll in the first two squares. Cross out the smaller number.
3. Reroll the die that shows the smaller number.
4. Record your second roll in the third square.
5. Record the sum of the two dice on the line.
6. The player with the larger sum wins the round.
7. Play 8 rounds. The player who wins the most rounds is the winner.

Fishing for 10

Materials ☐ number cards 0–10 (4 of each)
☐ *Fishing for 10* Record Sheet
Players 2–4
Skill Practicing combinations of 10
Object of the Game To collect as many combinations of 10 as you can.

Directions

1. Shuffle the cards.
2. Put the cards number-side down in a pile.
3. Each player takes 5 cards.
4. On your first turn, remove pairs of cards that add to 10 from your hand.
   Place them number-side up on the table.
   Draw cards from the pile so you have 5 cards again.
5. Take turns. When it is your turn again:
   - *Fish* by asking any other player for a card you need to make a combination of 10.

- If that player does not have the card you need, *go fish,* or draw a card from the pile.
- Place any combinations of 10 in your hand number-side up on the table.

6. Make sure everyone's cards add to 10.
7. After every round, draw cards from the pile if you have fewer than 5 cards in your hand.
8. Play until there are no more cards in the pile and nobody can make another combination of 10.
9. Record 4 of your combinations of 10 on your *Fishing for 10* Record Sheet.

Monster Squeeze

Materials ☐ 2 monster brackets
☐ number line

Players 4 or more

Skill Comparing numbers

Object of the Game To find the mystery number.

Directions

1. Choose roles. You need:
   - 1 leader
   - 2 monster movers
   - 1 or more players
2. The leader thinks of a mystery number.
3. The leader calls out 2 numbers.
   - One number is smaller than the mystery number.
   - One number is larger than the mystery number.
4. The monster movers cover these two numbers.

5. One player guesses a number between the monsters.
6. The leader tells whether the mystery number is larger or smaller than the guessed number.
7. One of the monster movers covers the guessed number.
8. Players keep guessing until they guess the mystery number.

Name that Number

Materials ☐ number cards 0–20 (4 of each card 0–10, and 1 of each card 11–20)

Players 2 to 4 (the game is more fun when played by 3 or 4 players)

Skill Using addition and subtraction to name equivalent numbers

Object of the Game To collect the most cards.

Directions

1. Shuffle the deck and place 5 cards number-side up on the table. Leave the rest of the deck number-side down. Then turn over the top card of the deck and lay it down next to the deck. The number on this card is the number to be named. Call this number the *target number.*

2. Players take turns. When it is your turn:
   - Try to name the target number by adding or subtracting the numbers on 2 or more of the 5 cards that are number-side up. A card may be used only once for each turn.

- If you can name the target number, take the cards you used to name it. Also take the target-number card. Then replace all the cards you took by drawing from the top of the deck.
- If you cannot name the target number, your turn is over. Turn over the top card of the deck and lay it down on the target-number pile. The number on this card is the new target number.

3. Play continues until all of the cards in the deck have been turned over. The player who has taken the most cards wins.

Mae and Joe take turns.

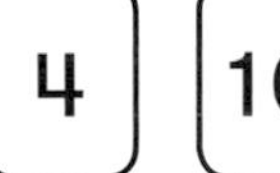

8 | 12 | 2

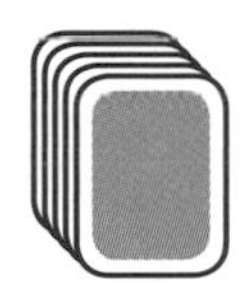

6

It is Mae's turn. The target number is 6. Mae names the number with $12 - 4 - 2$. She also could have said $4 + 2$ or $8 - 2$.

Mae takes the 12, 4, 2, and 6 cards. Then she replaces them by drawing cards from the deck. Now it is Joe's turn.

Number-Grid Difference

Materials
- ☐ number cards 0–9 (4 of each)
- ☐ 1 completed number grid
- ☐ 1 *Number-Grid Difference* Record Sheet
- ☐ 2 counters
- ☐ 1 calculator

Players 2

Skill Subtraction of 2-digit numbers using the number grid

Object of the Game To have the lower sum.

Directions

1. Shuffle the cards. Place the deck number-side down on the table.
2. Take turns. When it is your turn:
   - Take 2 cards from the deck and use them to make a 2-digit number. Place a counter on the grid to mark your number.
   - Find the difference between your number and your partner's number.
   - This difference is your score for the turn. Write the 2 numbers and your score on the Record Sheet.

3. Continue playing until each player has taken 5 turns and recorded 5 scores.
4. Find the sum of your 5 scores. You may use a calculator to add.
5. The player with the lower sum wins the game.

Ellie draws two 4s. She makes the number 44 and records it as her number on the Record Sheet.

Carlos draws a 6 and a 3 and makes the number 63. Ellie records 63 on the Record Sheet. Then Ellie subtracts.

$63 - 44 = 19$

Ellie records 19 as the difference.

Number-Grid Difference Record Sheet

Round	My Number	My Partner's Number	Difference (Score)
1	44	63	19
2			

Penny Plate

Materials ☐ 10 pennies
☐ 1 plate

Players 2

Skill Find pairs of numbers that add to 10

Object of the Game To get 5 points.

Directions

1. Player 1:
   - Turns the plate upside-down.
   - Hides some of the pennies under the plate.
   - Puts the remaining pennies on top of the plate.

2. Player 2:
   - Counts the pennies on top of the plate.
   - Figures out how many pennies are hidden under the plate.

3. If the number is correct, Player 2 gets a point.
4. Players trade roles and repeat Steps 1 and 2.
5. Each player keeps a tally of his or her points. The first player to get 5 points is the winner.

Another Way to Play

Use a different number of pennies.

Roll and Record Doubles

Materials ☐ *Roll and Record Doubles* Record Sheet
☐ 1 six-sided die

Players 2

Skill Finding addition doubles

Object of the Game To fill one column.

Directions

Work with a partner. When it is your turn:

1. Roll the die. Use that number to make a doubles fact.
For example:

$3 + 3 = 6$

2. Shade the first empty box above the sum for the doubles fact.
Take turns until one column is filled.

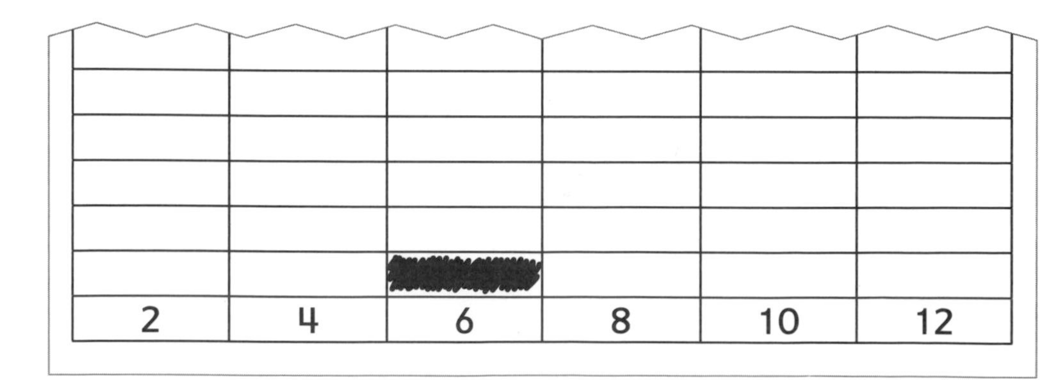

Subtraction Bingo

Materials ☐ number cards 0–10 (4 of each)
☐ *Subtraction Bingo* Game Mats
☐ counters

Players 2

Skill Subtraction facts 0–10

Object of the Game To get four in a row.

Directions

1. Shuffle the cards. Place the deck number-side down on the table.
2. Take turns. When it is your turn:
   - Flip over two cards and subtract. Call out the difference.
   - Your partner checks your answer.
   - You and your partner both cover one space on your game mats showing the difference.
3. Continue playing until one player covers 4 spaces in a row. That player calls out "Bingo!" and is the winner.

Salute!

Materials ☐ number cards 0–10 (4 of each)
Players 3
Skill Practicing addition and subtraction facts
Object of the Game To solve for the number on your card.

Directions

1. One person begins as the Dealer. The Dealer gives one card to each of the other two Players.
2. Without looking at their cards, the Players hold them on their foreheads with the number facing out.
3. The Dealer looks at both cards and says the sum of the two numbers.
4. Each Player looks at the other Player's card. They use the number they see and the sum said by the Dealer to figure out what the number on their card must be. They say that number out loud.

5. Once both Players have said their numbers, they can look at their own cards.
6. Rotate roles and repeat the game.
7. Play continues until everyone has been the Dealer five times, or until the entire deck of cards has been used.

The players use the number they see and the sum to try to figure out what their own number is.

Stop and Go

Materials
- ☐ 6 GO Cards (+ 9, + 8, + 7, + 6, + 10, + 20)
- ☐ 6 STOP Cards (− 0, − 0, − 10, − 10, − 20, − 20)
- ☐ *Stop and Go* Record Sheet

Players 2

Skill Adding and subtracting 2-digit numbers

Object of the Game To get to 50 (or to stop the player from getting to 50).

Directions

1. This game has two players:
   - The GO player
   - The STOP player
2. The GO player puts the GO cards number-side down and takes 1 GO card.
3. The GO player adds 20 to the amount on the GO card and records the sum on the *Stop and Go* Record Sheet.
4. The STOP player puts the STOP cards number-side down and takes 1 STOP card.

5. The STOP player subtracts the amount on the STOP card from the sum on the *Stop and Go* Record Sheet.

6. The players take turns adding and subtracting.
   - If the GO player reaches 50, the GO player wins.
   - If the STOP player pushes the GO player back to 0, the STOP player wins.
   - If the players run out of cards before the GO player reaches 50, the STOP player wins.

The Target Game

Materials
- ☐ number cards 0–9 (4 of each)
- ☐ base-10 blocks (10 longs and 30 cubes)
- ☐ 1 *Target Game* Mat for each player
- ☐ 1 *Target Game* Record Sheet for each player

Players 2

Skill Place value for whole numbers

Object of the Game To have 5 longs on the *Target Game* Mat.

Directions

1. Shuffle the number cards. Place the deck number-side down.
2. Players take turns. When it is your turn:
   - Turn over 2 cards. You may use either card to make a 1-digit number. Or, you may use both cards to make a 2-digit number.
   - Use base-10 blocks to model your number. Put these blocks just below your *Target Game* Mat, but not on the mat.

- You now have 2 choices:

Choice 1: You can add all of the base-10 blocks below the mat to the blocks already on your *Target Game* Mat.

Choice 2: You can subtract blocks equal in value to the base-10 blocks below the mat from the blocks already on your *Target Game* Mat. If you decide to subtract, you may first have to make exchanges on the *Target Game* Mat.

3. Players can make exchanges on their *Target Game* Mats at any time.

4. Play continues until the blocks on one player's mat have a value of 50 and show 5 longs. That player is the winner.

Alex was able to reach the target value of 50 in three turns:

Turns	Cards	Number Made	Addition or Subtraction on *Target Game* Mat	Value on Mat
1	6, 5	56	**Add** 5 longs and 6 cubes.	56
2	8, 9	8	Exchange 1 long for 10 cubes. **Subtract** 8 cubes.	48
3	5, 2	2	**Add** 2 cubes. Exchange 10 cubes for 1 long.	50

Tric-Trac

Materials ☐ 2 six-sided dice
☐ 22 pennies
☐ 1 *Tric-Trac* Game Mat for each player

Players 2

Skill Addition facts 0–10

Object of the Game To have the lower sum.

Directions

1. Cover the empty circles on your game mat with pennies.
2. Take turns. When it is your turn:
   - Roll the dice. Find the total number of dots. This is your sum.
   - Move 1 of your pennies and cover your sum on your game mat.

   OR

   - Move 2 or more of your pennies and cover any numbers that can be added together to equal your sum.

3. Play continues until no more numbers can be covered on your game mat. Your partner may continue playing, even after you are finished.

4. The game is over when neither player can cover any more numbers on his or her game mat.

5. Find the sum of your uncovered numbers. The player with the lower sum wins.

Set-up *Tric-Trac* Game Mat

Tric-Trac

0 1 2 3 4 5 6 7 8 9 10

Top-It

Materials ☐ number cards 0–15 (2 of each)
Players 2 or more
Skill Comparing numbers
Object of the Game To collect more cards.

Directions

1. Shuffle the cards. Place the deck number-side down on the table.
2. Each player turns over 1 card and says the number on it.
3. The player with the larger number takes all the cards. If two cards show the same number, those players turn over another card. The player with the larger number then takes all the cards for that round.
4. The game is over when all of the cards have been turned over.
5. The player with the most cards wins.

Other Ways to Play

Use Dominoes:

- Each player turns over 1 domino and says the total number of dots.
- The player with the larger number of dots takes both dominoes.
- The player with the most dominoes wins.

Use <, >, and = Cards:

- After each player turns over a card, put the <, >, or = card in between the cards to make a true number sentence. Read the number sentence out loud.
- The player with the larger number takes both number cards.
- The player with the most cards wins.

Make Large Numbers:

- Use only number cards 0–9 (4 of each). Get a Place-Value Mat.

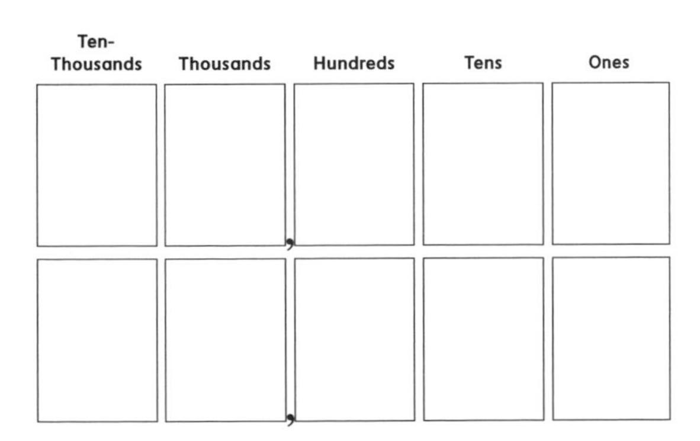

- Players decide on the number of digits they will use.
- Take turns turning over the top card and placing it on any of the empty boxes. When players have made a number using the number of digits they agreed on, they read the numbers aloud and compare them.
- The player with the largest number scores 1 point, the player with the next-largest number scores 2 points, and so on.
- The player with the fewest number of points after 5 rounds wins.

Practice Addition or Subtraction Facts:

- Use only number cards 0–9 (4 of each). Choose to practice addition or subtraction facts.
- Each player turns over 2 cards and calls out the sum or difference. The player with the largest sum or difference takes all the cards.
- In the case of a tie, each player turns over 2 more cards and calls out the sum or difference. The player with the largest sum or difference takes all of the cards.
- Play until there are no cards remaining. The player with the most cards wins.

Technology

Calculators and Computers

Read It Together

A **calculator** is a tool that can count, add, subtract, multiply, and divide. Not all calculators are alike. Some calculators are on computers or phones.

Here is one type of calculator.

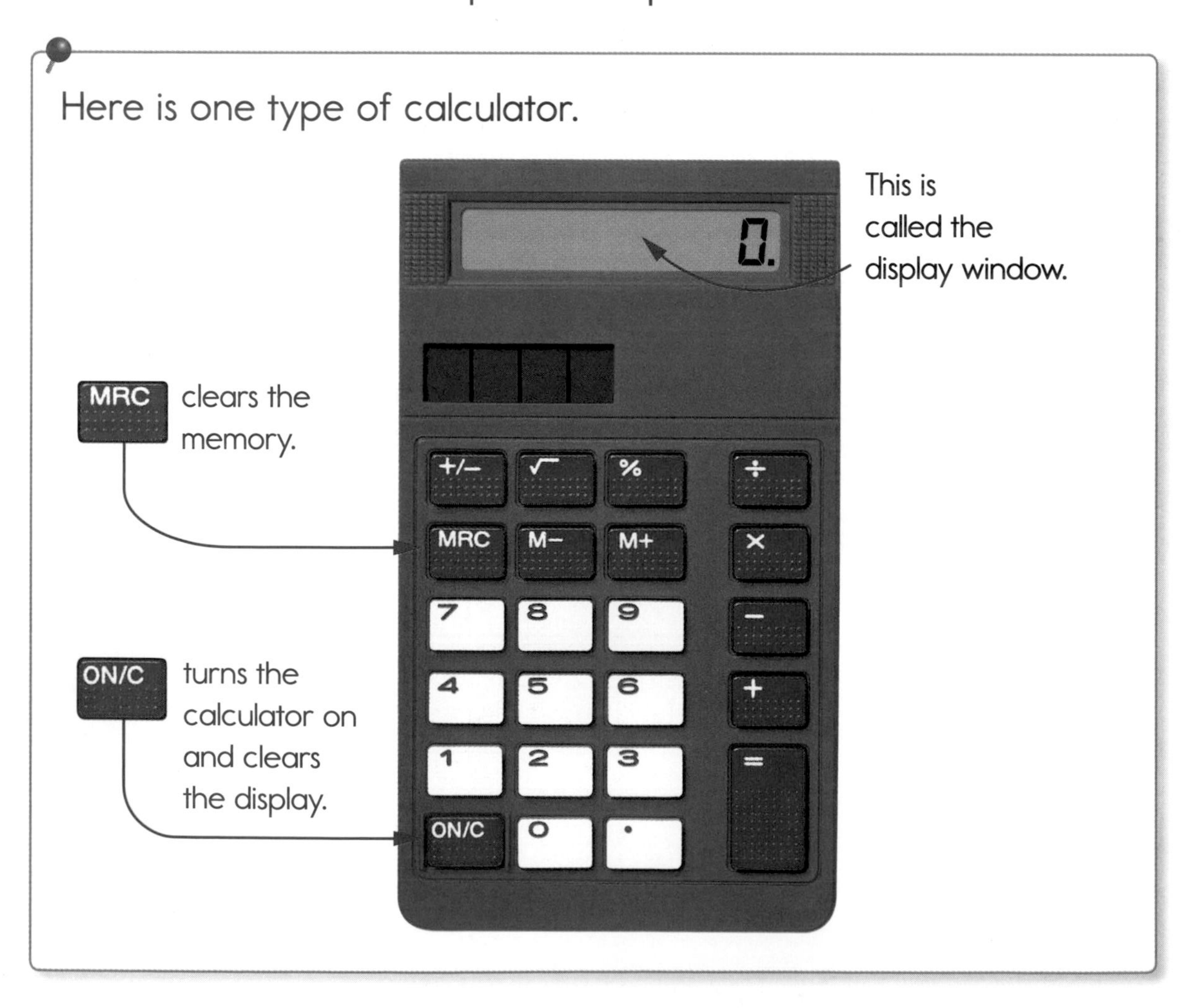

Calculators can help you solve problems that have large numbers or many numbers. You can use a calculator to find correct answers quickly.

Note When you use a calculator, estimate to check whether the answer makes sense.

You can **skip count** up or back on a calculator.

Use [calculator] to skip count. Start at 1. Count up by 2s.

Program the Calculator	Keys to Press	Display
Clear the display.	ON/C	0.
Enter the starting number.	1	1.
Tell it to count up by a number.	+ 2	2.
Tell it to skip to the next number.	=	3.
Tell it to skip to the next number.	=	5.
Tell it to skip to the next number.	=	7.

You can count back by pressing [−] instead of [+].

You can use **computers** to learn about mathematics and find real-world data on the Internet.

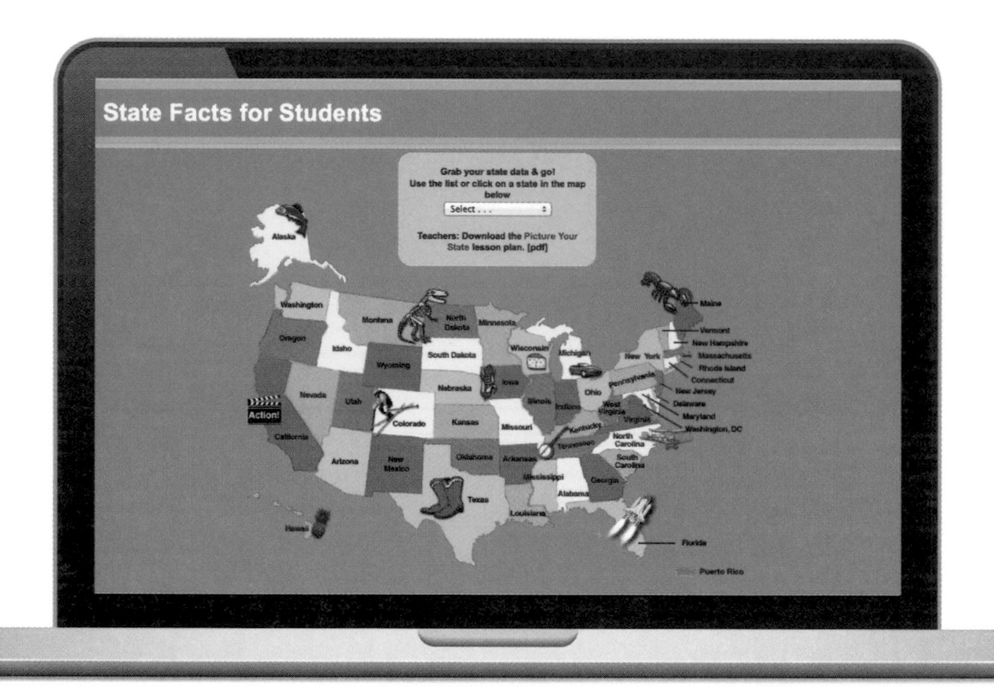

You can use small computers, such as tablets and smartphones, to do and explore mathematics.

You can

- solve problems
- investigate shapes
- make tables and graphs
- play math games

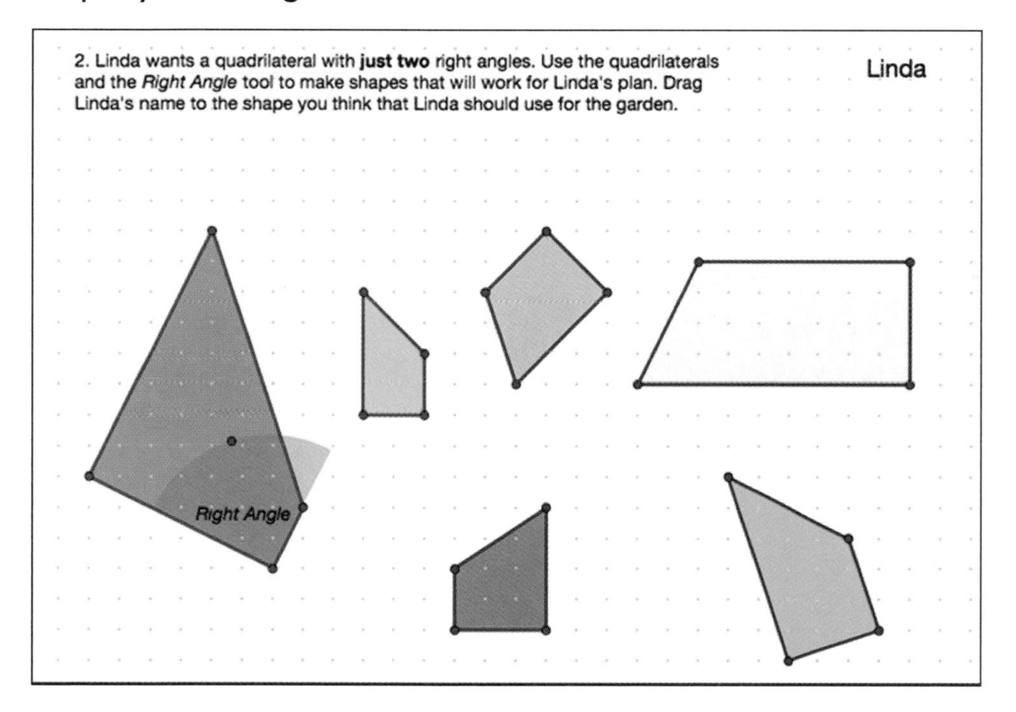

Computers let you save your work and share it with teachers, family, and other children.

A

B

C

MRB
179